Communications
in Computer and Information Science   1316

Commenced Publication in 2007
Founding and Former Series Editors:
Simone Diniz Junqueira Barbosa, Phoebe Chen, Alfredo Cuzzocrea,
Xiaoyong Du, Orhun Kara, Ting Liu, Krishna M. Sivalingam,
Dominik Ślęzak, Takashi Washio, Xiaokang Yang, and Junsong Yuan

More information about this series at http://www.springer.com/series/7899

Vesna Dimitrova · Ivica Dimitrovski (Eds.)

ICT Innovations 2020

Machine Learning and Applications

12th International Conference, ICT Innovations 2020
Skopje, North Macedonia, September 24–26, 2020
Proceedings

 Springer

Editors
Vesna Dimitrova
University Ss. Cyril and Methodius
Skopje, North Macedonia

Ivica Dimitrovski
University Ss. Cyril and Methodius
Skopje, North Macedonia

ISSN 1865-0929   ISSN 1865-0937   (electronic)
Communications in Computer and Information Science
ISBN 978-3-030-62097-4   ISBN 978-3-030-62098-1   (eBook)
https://doi.org/10.1007/978-3-030-62098-1

This Springer imprint is published by the registered company Springer Nature Switzerland AG
The registered company address is: Gewerbestrasse 11, 6330 Cham, Switzerland

Preface

The ICT Innovations conference series, organized by the Macedonian Society of Information and Communication Technologies (ICT-ACT) is one of the leading international conferences in the region that serves as a platform for presenting novel ideas and fundamental advances in the fields of computer science and engineering. The 12th ICT Innovations 2020 conference that brought together academics, students, and industrial practitioners, was held online from September 24–26, 2020.

The focal point for this year's conference was machine learning and applications. The need for machine learning is ever growing due to the increased pervasiveness of data analysis tasks in almost every area of life, including business, science, and technology. Not only is the pervasiveness of data analysis tasks increasing, but so is their complexity. These tasks usually need to handle massive datasets which can be partially labeled. The datasets can have many input and output dimensions, streaming at very high rates. Furthermore, the data can be placed in a spatio-temporal or network context.

Machine learning is continuously unleashing its power in a wide range of applications including marketing, e-commerce, software systems, networking, telecommunications, banking, finance, economics, social science, computer vision, speech recognition, natural language processing, robotics, biology, transportation, health care, and medicine. The information explosion has resulted in the collection of massive amounts of data. This amount of data, coupled with the rapid development of processor power and computer parallelization, has made it possible to build different predictive models. These predictive models can study and analyze complex data in various application areas to discover relationships in data to inspire insights and create opportunities, identify anomalies and solve problems, anticipate outcomes, and make better decisions.

Some of these topics were brought to the forefront of the ICT Innovations 2020 conference. This book presents a selection of papers presented at the conference which contributed to the discussions on various aspects of machine learning and applications. The conference gathered 146 authors from 14 countries reporting their scientific work and solutions in ICT. Only 18 papers were selected for this edition by the International Program Committee, consisting of 163 members from 45 countries, chosen for their scientific excellence in their specific fields.

We would like to express our sincere gratitude to the invited speakers for their inspirational talks, to the authors for submitting their work and sharing their most recent research, and the reviewers for providing their expertize during the selection process. Special thanks go to Bojana Koteska and Monika Simjanovska for the technical preparation of the conference proceedings.

September 2020

Vesna Dimitrova
Ivica Dimitrovski

Organization

Conference and Program Chairs

Vesna Dimitrova   Ss. Cyril and Methodius University, North Macedonia
Ivica Dimitrovski   Ss. Cyril and Methodius University, North Macedonia

Program Committee

Jugoslav Achkoski   General Mihajlo Apostolski Military Academy, North Macedonia
Nevena Ackovska   Ss. Cyril and Methodius University, North Macedonia
Syed Ahsan   Technische Universität Graz, Austria
Marco Aiello   University of Groningen, The Netherlands
Zahid Akhtar   University of Memphis, USA
Luis Alvarez Sabucedo   Universidade de Vigo, Spain
Ljupcho Antovski   Ss. Cyril and Methodius University, North Macedonia
Goce Armenski   Ss. Cyril and Methodius University, North Macedonia
Hrachya Astsatryan   National Academy of Sciences of Armenia, Armenia
Tsonka Baicheva   Bulgarian Academy of Science, Bulgaria
Verica Bakeva   Ss. Cyril and Methodius University, North Macedonia
Valentina Emilia Balas   Aurel Vlaicu University of Arad, Romania
Antun Balaz   Institute of Physics Belgrade, Serbia
Lasko Basnarkov   Ss. Cyril and Methodius University, North Macedonia
Slobodan Bojanic   Universidad Politécnica de Madrid, Spain
Dragan Bosnacki   Eindhoven University of Technology, The Netherlands
Torsten Braun   University of Bern, Switzerland
Andrej Brodnik   University of Ljubljana, Slovenia
Serhat Burmaoglu   Izmir Katip Çelebi University, Turkey
Francesc Burrull   Universidad Politécnica de Cartagena, Spain
Ioanna Chouvarda   Aristotle University of Thessaloniki, Greece
Betim Cico   Epoka University, Albania
Emmanuel Conchon   Institut de Recherche en Informatique de Toulouse, France
Ivan Corbev   Ss. Cyril and Methodius University, North Macedonia
Marilia Curado   University of Coimbra, Portugal
Robertas Damasevicius   Kaunas University of Technology, Lithuania
Ashok Kumar Das   IIIT Hyderabad, India
Danco Davcev   Ss. Cyril and Methodius University, North Macedonia
Antonio De Nicola   ENEA, Italy
Aleksandra Dedinec   Ss. Cyril and Methodius University, North Macedonia
Domenica D'Elia   Institute for Biomedical Technologies, Italy
Boris Delibašić   University of Belgrade, Serbia

Vesna Dimitrievska Ristovska	Ss. Cyril and Methodius University, North Macedonia
Vesna Dimitrova	Ss. Cyril and Methodius University, North Macedonia
Ivica Dimitrovski	Ss. Cyril and Methodius University, North Macedonia
Salvatore Distefano	University of Messina, Italy
Milena Djukanovic	University of Montenegro, Montenegro
Martin Drlik	Constantine the Philosopher University, Slovakia
Saso Dzeroski	Jožef Stefan Institute, Slovenia
Tome Eftimov	Stanford University, USA, and Jožef Stefan Institute, Slovenia
Suliman Mohamed Fati	INTI International University, Malaysia
Christian Fischer Pedersen	Aarhus University, Denmark
Kaori Fujinami	Tokyo University of Agriculture and Technology, Japan
Slavko Gajin	University of Belgrade, Serbia
Ivan Ganchev	University of Limerick, Ireland, and Plovdiv University, Bulgaria
Todor Ganchev	Technical University of Varna, Bulgaria
Amjad Gawanmeh	University of Dubai, UAE
Ilche Georgievski	University of Groningen, The Netherlands
Sonja Gievska	Ss. Cyril and Methodius University, North Macedonia
Hristijan Gjoreski	Ss. Cyril and Methodius University, North Macedonia
Dejan Gjorgjevikj	Ss. Cyril and Methodius University, North Macedonia
Rossitza Goleva	Technical University of Sofia, Bulgaria
Abel Gomes	University of Beira Interior, Portugal
Sasho Gramatikov	Ss. Cyril and Methodius University, North Macedonia
Andrej Grgurić	Ericsson Nikola Tesla, Croatia
David Guralnick	International E-Learning Association, France
Marjan Gushev	Ss. Cyril and Methodius University, North Macedonia
Elena Hadzieva	University of Information Science and Technology, St. Paul the Apostle, North Macedonia
Violeta Holmes	University of Huddersfield, UK
Fu Shiung Hsieh	Chaoyang University of Technology, Taiwan
Ladislav Huraj	Ss. Cyril and Methodius University, Slovakia
Hieu Trung Huynh	Industrial University of Ho Chi Minh City, Vietnam
Sergio Ilarri	University of Zaragoza, Spain
Natasha Ilievska	Ss. Cyril and Methodius University, North Macedonia
Ilija Ilievski	Graduate School for Integrative Sciences and Engineering, Singapore
Boro Jakimovski	Ss. Cyril and Methodius University, North Macedonia
Smilka Janeska-Sarkanjac	Ss. Cyril and Methodius University, North Macedonia
Metodija Jancheski	Ss. Cyril and Methodius University, North Macedonia
Aleksandar Jevremović	Singidunum University, Serbia
Mile Jovanov	Ss. Cyril and Methodius University, North Macedonia
Milos Jovanovik	Ss. Cyril and Methodius University, North Macedonia
Vacius Jusas	Kaunas University of Technology, Lithuania

Slobodan Kalajdziski	Ss. Cyril and Methodius University, North Macedonia
Kalinka Kaloyanova	University of Sofia, Bulgaria
Ivan Kitanovski	Ss. Cyril and Methodius University, North Macedonia
Mirjana Kljajic Borstnar	University of Maribor, Slovenia
Dragi Kocev	Jožef Stefan Institute, Slovenia
Margita Kon-Popovska	Ss. Cyril and Methodius University, North Macedonia
Magdalena Kostoska	Ss. Cyril and Methodius University, North Macedonia
Bojana Koteska	Ss. Cyril and Methodius University, North Macedonia
Petra Kralj Novak	Jožef Stefan Institute, Slovenia
Andrea Kulakov	Ss. Cyril and Methodius University, North Macedonia
Arianit Kurti	Linnaeus University, Sweden
Petre Lameski	Ss. Cyril and Methodius University, North Macedonia
Sanja Lazarova-Molnar	University of Southern Denmark, Denmark
Hwee San Lim	University of Science, Malaysia
Suzana Loshkovska	Ss. Cyril and Methodius University, North Macedonia
José Machado Da Silva	University of Porto, Portugal
Ana Madevska Bogdanova	Ss. Cyril and Methodius University, North Macedonia
Gjorgji Madjarov	Ss. Cyril and Methodius University, North Macedonia
Ninoslav Marina	University of Science and Technology, St. Paul the Apostole, North Macedonia
Smile Markovski	Ss. Cyril and Methodius University, North Macedonia
Marcin Michalak	Silesian University of Technology, Poland
Hristina Mihajloska	Ss. Cyril and Methodius University, North Macedonia
Aleksandra Mileva	Goce Delčev University of Štip, North Macedonia
Biljana Mileva Boshkoska	Faculty of Information Studies in Novo Mesto, Slovenia
Georgina Mirceva	Ss. Cyril and Methodius University, North Macedonia
Miroslav Mirchev	Ss. Cyril and Methodius University, North Macedonia
Igor Mishkovski	Ss. Cyril and Methodius University, North Macedonia
Kosta Mitreski	Ss. Cyril and Methodius University, North Macedonia
Pece Mitrevski	University St. Clement of Ohrid, North Macedonia
Irina Mocanu	University POLITEHNICA of Bucharest, Romania
Ammar Mohammed	Cairo University, Egypt
Andreja Naumoski	Ss. Cyril and Methodius University, North Macedonia
Ivana Ognjanovic	University of Donja Gorica, Montenegro
Pance Panov	Jožef Stefan Institute, Slovenia
Marcin Paprzycki	Polish Academy of Sciences, Poland
Dana Petcu	West University of Timisoara, Romania
Konstantinos Petridis	Hellenic Mediterranean University, Greece
Matus Pleva	Technical University of Košice, Slovakia
Vedran Podobnik	University of Zagreb, Croatia
Florin Pop	University POLITEHNICA of Bucharest, Romania
Zaneta Popeska	Ss. Cyril and Methodius University, North Macedonia
Aleksandra Popovska-Mitrovikj	Ss. Cyril and Methodius University, North Macedonia
Marco Porta	University of Pavia, Italy

Rodica Potolea	Technical University of Cluj-Napoca, Romania
Ustijana Rechkoska Shikoska	University of Science and Technology, North Macedonia
Manjeet Rege	University of St. Thomas, USA
Miriam Reiner	Technion - Israel Institute of Technology, Israel
Pance Ribarski	Ss. Cyril and Methodius University, North Macedonia
Blagoj Ristevski	University St. Clement of Ohrid, North Macedonia
Sasko Ristov	University of Innsbruck, Austria
David Šafránek	Masaryk University, Czech Republic
Jatinderkumar Saini	Narmada College of Computer Application, India
Simona Samardjiska	Radboud University, The Netherlands
Snezana Savoska	University St. Clement of Ohrid, North Macedonia
Loren Schwiebert	Wayne State University, USA
Bryan Scotney	Ulster University, UK
Gjorgji Strezoski	University of Amsterdam, The Netherlands
Osman Ugur Sezerman	Acibadem University, Turkey
Vladimir Siládi	Matej Bel University, Slovakia
Josep Silva	Universitat Politècnica de València, Spain
Manuel Silva	INESC TEC, Portugal
Nikola Simidjievski	Jožef Stefan Institute, Slovenia
Monika Simjanoska	Ss. Cyril and Methodius University, North Macedonia
Ana Sokolova	University of Salzburg, Austria
Michael Sonntag	Johannes Kepler University Linz, Austria
Dejan Spasov	Ss. Cyril and Methodius University, North Macedonia
Riste Stojanov	Ss. Cyril and Methodius University, North Macedonia
Milos Stojanovic	College of Applied Technical Sciences in Nis, Serbia
Dario Stojanovski	LMU Munich, Germany
Biljana Stojkoska	Ss. Cyril and Methodius University, North Macedonia
Stanimir Stoyanov	Plovdiv University, Bulgaria
Biljana Tojtovska	Ss. Cyril and Methodius University, North Macedonia
Dimitar Trajanov	Ss. Cyril and Methodius University, North Macedonia
Ljiljana Trajkovic	Simon Fraser University, Canada
Vladimir Trajkovik	Ss. Cyril and Methodius University, North Macedonia
Denis Trcek	University of Ljubljana, Slovenia
Christophe Trefois	University of Luxembourg, Luxembourg
Kire Trivodaliev	Ss. Cyril and Methodius University, North Macedonia
Katarina Trojacanec	Ss. Cyril and Methodius University, North Macedonia
Zlatko Varbanov	University of Veliko Tarnovo, Bulgaria
Goran Velinov	Ss. Cyril and Methodius University, North Macedonia
Mitko Veta	Eindhoven University of Technology, The Netherlands
Elena Vlahu-Gjorgievska	University of Wollongong, Australia
Boris Vrdoljak	University of Zagreb, Croatia
Wibowo Santoso	Central Queensland University, Australia
Shuxiang Xu	University of Tasmania, Australia
Malik Yousef	Zefat Academic College, Israel
Wuyi Yue	Konan University, Japan

Eftim Zdravevski	SS. Cyril and Methodius University, North Macedonia
Katerina Zdravkova	Ss. Cyril and Methodius University, North Macedonia
Jurica Zucko	Faculty of Food Technology and Biotechnology, Croatia

Scientific Committee

Danco Davcev	Ss. Cyril and Methodius University, North Macedonia
Dejan Gjorgjevikj	Ss. Cyril and Methodius University, North Macedonia
Boro Jakimovski	Ss. Cyril and Methodius University, North Macedonia
Aleksandra Popovska-Mitrovikj	Ss. Cyril and Methodius University, North Macedonia
Vesna Dimitrova	Ss. Cyril and Methodius University, North Macedonia
Ivica Dimitrovski	Ss. Cyril and Methodius University, North Macedonia

Technical Committee

Bojana Koteska	Ss. Cyril and Methodius University, North Macedonia
Monika Simjanovska	Ss. Cyril and Methodius University, North Macedonia
Kostadin Mishev	Ss. Cyril and Methodius University, North Macedonia
Stefan Andonov	Ss. Cyril and Methodius University, North Macedonia
Vlatko Spasev	Ss. Cyril and Methodius University, North Macedonia
Nasi Jofce	Ss. Cyril and Methodius University, North Macedonia

Keynote Abstracts

Generic Acceleration Schemes for Large-Scale Optimization in Machine Learning

Julien Mairal

University of Grenoble Alpes, Inria, CNRS, Grenoble INP, LJK,
38000 Grenoble, France
julien.mairal@inria.fr

Abstract. In this talk, we will present a few optimization principles that have been shown to be useful to address large-scale problems in machine learning. We will focus on recent variants of the stochastic gradient descent method that benefit from several acceleration mechanisms such as variance reduction and Nesterov's extrapolation. We will discuss both theoretical results in terms of complexity analysis, and practical deployment of these approaches, demonstrating that even though Nesterov's acceleration method is almost 40 years old, it is still highly relevant today.

Keywords: Optimization algorithms · Nesterov's extrapolation · Stochastic gradient descent

A Decade of Machine Learning in Profiled Side-Channel Analysis

Stjepan Picek

Delft University of Technology (TU Delft), Netherlands
S.Picek@tudelft.nl

Abstract. The growing markets of embedded computing, and especially Internet-of-Things, require large amounts of confidential data to be processed on electronic devices. Cryptographic algorithms are usually implemented as part of those systems and, if not adequately protected, are vulnerable to side-channel analysis (SCA). In SCA, the attacker exploits weaknesses in the physical implementations of cryptographic algorithms. SCA can be divided into profiled and non-profiled analysis. The profiled side-channel analysis considers a scenario where the adversary has control over a device identical to the target device. The adversary then learns statistics from the device under control and tries to match them on other target devices. In the last decade, profiled side-channel analysis based on machine learning proved to be very successful in breaking cryptographic implementations in various settings. Despite successful attacks, even in the presence of countermeasures, there are many open questions. A large part of the research concentrates on improving the performance of attacks. At the same time, little is done to understand them and, even more importantly, use that knowledge to design more secure implementations. In this talk, we start by discussing success stories on machine learning-based side-channel analysis. First, we discuss how to improve deep learning-based SCA performance by using ensembles of neural networks. Next, we concentrate on the pre-processing phase and explore whether the countermeasures could be considered noise and removed before the attack. More precisely, we investigate the performance of denoising autoencoders to remove several types of hiding countermeasures. Finally, we discuss how to use the results of non-profiled methods as a start for the deep learning-based analysis. To be able to run such an analysis, we discuss the iterative deep learning-based SCA framework for public-key cryptography. In the second part of this talk, we concentrate on critical open questions and research directions that still need to be explored. Here, we will discuss questions like the difference between machine learning and SCA metrics, or portability (the differences between the training and testing devices). Finally, we briefly connect the machine learning progress in profiled SCA with developments in other security domains (hardware Trojans, fault injection attacks).

Keywords: Side-channel analysis · Profiled Attacks · Machine learning · Deep learning

The Machine Learning of Time and Applications

Efstratios Gavves

University of Amsterdam
egavves@uva.nl

Abstract. Visual artificial intelligence automatically interprets what happens in visual data like videos. Today's research strives with queries like: "Is this person playing basketball?"; "Find the location of the brain stroke"; or "Track the glacier fractures in satellite footage". All these queries are about visual observations already taken place. Today's algorithms focus on explaining past visual observations. Naturally, not all queries are about the past: "Will this person draw something in or out of their pocket?"; "Where will the tumour be in 5 seconds given breathing patterns and moving organs?"; or, "How will the glacier fracture given the current motion and melting patterns?". For these queries and all others, the next generation of visual algorithms must expect what happens next given past visual observations. Visual artificial intelligence must also be able to prevent before the fact, rather than explain only after it. In this talk, I will present my vision on what these algorithms should look like and how we can obtain them. Furthermore, I will present some recent works and applications in this direction within my lab and spinoff.

Keywords: Visual artificial intelligence · Action recognition · Online action detection

Deep Learning for Machinery Fault Diagnosis

Chuan Li

School of Mechanical Engineering, Dongguan University of Technology, China
chuanli@dgut.edu.cn

Abstract. Machinery fault diagnosis aims at diagnosing the causes of the degradation of the industrial machinery and assessing the degradation level, with the objective of increasing the system availability and reducing operation and maintenance costs. In the era of Industry 4.0, the increased availability of information from industrial monitored machinery and the grown ability of treating the acquired information by intelligent algorithms have opened the wide doors for the development of advanced diagnosis methods. However, traditional methods highly rely on the design and selection of handcrafted features, which require the knowledge of experts and the use of computationally intensive trial and error approaches. The large amount of available data poses the problem of extracting and selecting the features relevant for the development of the fault diagnosis methods. Deep learning is a promising tool for dealing with this problem. Deep learning refers to a class of methods that are capable of extracting hierarchical representations from huge volumes of large-dimensional data by using neural networks with multiple layers of non-linear transformations. Therefore, they are expected to directly and automatically provide high-level abstractions of the big data available in Industry 4.0, without requiring human-designed and labor-intensive analyses of the data for the extraction of degradation features. This talk aims at contributing to the above scenario by presenting advanced deep learning methods for modelling complex industrial machinery and treating their data, with the objective of diagnosing their failures.

Keywords: Deep learning · Big data · Fault diagnosis · Machinery · Intelligent algorithm

Contents

Temperature Dependent Initial Chemical Conditions for WRF-Chem
Air Pollution Simulation Model 1
    Nenad Anchev, Boro Jakimovski, Vlado Spiridonov, and Goran Velinov

Smart City Air Pollution Monitoring and Prediction:
A Case Study of Skopje. 15
    Jovan Kalajdjieski, Mladen Korunoski, Biljana Risteska Stojkoska,
    and Kire Trivodaliev

Cloud Based Personal Health Records Data Exchange in the Age of IoT:
The Cross4all Project. 28
    Savoska Snezana, Vassilis Kilintzis, Boro Jakimovski, Ilija Jolevski,
    Nikolaos Beredimas, Alexandros Mourouzis, Ivan Corbev,
    Ioanna Chouvarda, Nicos Maglaveras, and Vladimir Trajkovik

Time Series Anomaly Detection with Variational Autoencoder Using
Mahalanobis Distance 42
    Laze Gjorgiev and Sonja Gievska

Towards Cleaner Environments by Automated Garbage Detection
in Images. 56
    Aleksandar Despotovski, Filip Despotovski, Jane Lameski,
    Eftim Zdravevski, Andrea Kulakov, and Petre Lameski

Fine – Grained Image Classification Using Transfer Learning
and Context Encoding. 64
    Marko Markoski and Ana Madevska Bogdanova

Machine Learning and Natural Language Processing: Review of Models
and Optimization Problems. 71
    Emiliano Mankolli and Vassil Guliashki

Improving NER Performance by Applying Text Summarization
on Pharmaceutical Articles. 87
    Jovana Dobreva, Nasi Jofche, Milos Jovanovik, and Dimitar Trajanov

Case Study: Predicting Students Objectivity in Self-evaluation Responses
Using Bert Single-Label and Multi-Label Fine-Tuned
Deep-Learning Models. 98
    Vlatko Nikolovski, Dimitar Kitanovski, Dimitar Trajanov,
    and Ivan Chorbev

Fat Tree Algebraic Formal Modelling Applied to Fog Computing   111
   Pedro Juan Roig, Salvador Alcaraz, Katja Gilly, and Sonja Filiposka

A Circuit for Flushing Instructions from Reservation Stations
in Microprocessors   127
   Dejan Spasov

Parallel Programming Strategies for Computing Walsh Spectra
of Boolean Functions.   138
   Dushan Bikov and Maria Pashinska

Pipelined Serial Register Renaming.   153
   Dejan Spasov

Fast Decoding with Cryptcodes for Burst Errors   162
   Aleksandra Popovska-Mitrovikj, Verica Bakeva,
   and Daniela Mechkaroska

Cybersecurity Training Platforms Assessment.   174
   Vojdan Kjorveziroski, Anastas Mishev, and Sonja Filiposka

Real-Time Monitoring and Assessing Open Government Data:
A Case Study of the Western Balkan Countries   189
   Vigan Raça, Nataša Veljković, Goran Velinov, Leonid Stoimenov,
   and Margita Kon-Popovska

Analysis of Digitalization in Healthcare: Case Study   202
   Goce Gavrilov, Orce Simov, and Vladimir Trajkovik

Correlating Glucose Regulation with Lipid Profile.   217
   Ilija Vishinov, Marjan Gusev, Lidija Poposka, and Marija Vavlukis

Author Index   229

Temperature Dependent Initial Chemical Conditions for WRF-Chem Air Pollution Simulation Model

Nenad Anchev, Boro Jakimovski[(✉)], Vlado Spiridonov, and Goran Velinov

Faculty of Computer Science and Engineering, Ss. Cyril and Methodius University in Skopje, Skopje, North Macedonia
`boro.jakimovski@finki.ukim.mk`

Abstract. Air pollution is a health hazard that has been brought to public attention in the recent years, due to the widespread networks of air quality measurement stations. The importance of the problem brought the need to develop accurate air prediction models. The coupled meteo-chemical simulation systems have already been demonstrated to correctly predict the episodes of high pollution events. Due to the complexity of these models, which simulate the emissions, interactions and transport of pollutants in the atmosphere, setting up the correct parameters tailored for a specific area is a challenging task. In this paper we present an exhaustive analysis of the historical air pollution measurements, a detailed evaluation of an existing WRF-Chem based predictive model and propose an approach for improvement of that specific model. We use a specific temperature-dependent way of scaling the initial chemical conditions of a WRF-chem simulation, which leads to significant reduction of the bias by the model. We present the analysis that led us into these conclusions, the setup of the model, and the improvements made by using this approach.

Keywords: WRF-Chem · Air pollution · Prediction · Simulation · Initial conditions · PM10 modeling · Temperature dependent

1 Introduction

The problem of air pollution has been in the focus of the public interest in our country in the past few years. The enabling reason for that was the availability of air pollution monitoring stations with publicly available data in real time [1, 2]. The data measured are showing consistently high pollution with PM10 and PM2.5 during winter, often with high pollution events with measurements up to more than 10 times the limit of what is considered a safe concentration [3]. These episodes of extreme air pollution have increased the public interest in monitoring the pollution levels in real time, for the purpose of personal planning of outdoor activities and minimizing exposure to hazardous air.

A step towards in the direction of more effective planning and preparedness for minimizing exposure to hazardous air would be the creation of a reliable model for air

© Springer Nature Switzerland AG 2020
V. Dimitrova and I. Dimitrovski (Eds.): ICT Innovations 2020, CCIS 1316, pp. 1–14, 2020.
https://doi.org/10.1007/978-3-030-62098-1_1

pollution prediction. Few such systems have appeared in the recent years, with some of them still in the research phase, while others have been brought to the stage of user-accessible air quality prediction apps. These systems are based on a variety of different models, like WRF-Chem, SILAM, CAMS and others [4–8]. In this paper we are going to take a more detailed look in the WRF-Chem model and work on its improvement.

Most of these models, while working relatively correctly throughout most of Europe, struggle to correctly predict the pollution concentrations in the area of the Balkans. The discrepancy is mostly due to two factors, the first one is the complex orography, which often causes temperature inversions in the valleys of the larger towns, where localized extreme high concentrations of pollutants occur. The second one is the use of wood and other solid fuels for domestic space heating which is a large non-industrial pollution source in these areas, further amplified by the orographic effect. These domestic emissions are hard to be modeled properly and are endemic to these areas. The pan-European prediction models focusing mostly on industrial and transportation emissions often fail to integrate these domestic emissions properly on a local level. The combination of the two factors often leads to these models making incorrect predictions of the pollution in the larger urban areas in the Balkans.

The solution to the problem caused by orography is increasing the simulation grid density around the urban areas, and thus improve the localized predictions. This is a fairly straightforward task, but due to the significantly increased computational demand of the models, we are not be able to test it in our experiments for this paper.

The second problem is a more challenging one, and we will try to tackle it in our research. Our hypothesis is that the total pollution emissions during the winter have a strong component originating from the domestic heating. As we expect that the cumulative solid fuels combustion intensity (and therefore the pollutants emissions) depends on the air temperature, we propose a temperature dependent scaling of the pollution emissions. We will show that the improved accuracy of estimating the pollutants emissions will cause an overall improvement of the correctness of the model.

The base model we will be improving is a WRF-Chem model, which has already been developed for our country [9] We will present the setup of that model in Sect. 2, which we will refer to as 'the initial model'. In Sect. 3 we will analyze the data from air pollution measurements and try to get insight and find patterns that will define our model. In Sect. 4 we elaborate the improved model and present our contribution, based on the combined findings of the previous two chapters. Section 5 presents the results of the evaluation of the model, with a discussion on the obtained improvements. In the final section, we present our conclusion and further improvements possible in future.

2 WRF-Chem Model

The core meteo-chemical model setup we use is based on the WRF-Chem model. Its core is the Weather Research and Forecasting (WRF) Model, a mesoscale numerical weather prediction system designed for both atmospheric research and operational forecasting applications [10]. The WRF-Chem extension simulates the emission, transport, mixing, and chemical transformation of trace gases and aerosols simultaneously with the meteorology [11].

We use the setup of the simulation as proposed for a similar model [9], which we are going to describe in this section thoroughly. The initial and boundary meteorological conditions are prepared by WPS (WRF Preprocessing System) and the initial chemical conditions are calculated by PREP-CHEM-SRC [12]. A crucial part of their setup is the addition of user-defined data, where they provide the grid of yearly emissions on the entire territory of the Republic of North Macedonia, aggregated by economic sectors and sources. This grid of sector aggregated pollution sources is compiled by the MOEPP [13, 14], and with the addition of these data, they are able to correctly predict the episodes of high pollution in their evaluation domain.

The meteorological module of the model is based on the Advanced Research WRF (ARW) core developed by the NCAR [15]. ARW is a non-hydrostatic mesoscale model with a compressible equation on C-grid staggering, using a terrain following hydrostatic pressure-vertical coordinate. It conserves mass, momentum, dry entropy, and scalars using a flux-conserving form for all prognostic equations. The numerical methods utilize third-order Runge-Kutta split-explicit time differencing, together with higher-order advection.

The Weather Research and Forecasting model ARW is coupled with chemistry (WRF-Chem), as an efficient and flexible system for weather and air quality forecast. WRF-Chem was developed at NOAA/ESRL (National Oceanic and Atmospheric Administration/Earth System Research Laboratory) [16] and updated by incorporating complex gas-phase chemistry, aerosol treatments, and photolysis scheme [17]. The air quality component of WRF-Chem is fully consistent with the meteorological component; both components use the same transport scheme (mass and scalar preserving), the same horizontal and vertical grids, the same physical schemes for subgrid scale transport, and the same time step for transport and vertical mixing.

In the initial model, WRF-Chem v.4.0 released at NCEP in June 2018 is employed as a basis for the chemical transport forecast. The Regional Acid Deposition Model version 2 (RADM2) chemical mechanism for gas-phase chemistry schemes [18] is without kinetic pre-processor (KPP). In addition, the air quality modeling system includes also the Modal Aerosol Dynamics Model for Europe (MADE) [19], coupled with the SORGAM (Secondary Organic Aerosol Model) parameterization [20] for PM10 simulations.

The urban emissions are derived from daily inventories built in the emission preprocessor PREP-CHEM-SRC. The PREP-CHEM-SRC is a tool developed to estimate the emission fields of aerosols and trace gases from biomass burning (by satellite observations and inventories), biogenic, urban-industrial, biofuel use, and volcanic and agricultural waste burning sources for regional and global transport models based on available inventories and products [21]. The main objective of PREP-CHEM-SRC is to estimate the emission fields of the main trace gases and aerosols for use in atmospheric-chemistry transport models, such as WRF-CHEM.

In this work, we are using the global anthropogenic emission data for gaseous species (CO2, CO, NOx = NO + NO2, SO2, NH3) compiled and distributed by the Emission Database for Global Atmospheric Research (EDGAR) system (http://www.mnp.nl/edgar) [22]. The EDGAR-HTAP project compiled a global emission data set with annual inventories for CH4, NMVOC, CO, SO2, NOx, NH3, PM10, PM2.5, BC, and OC and covering the period 2000–2005 for 10 aggregated sectors and on a global $0.1° \times 0.1°$

resolution. The global emission data comes from the Reanalysis of the Tropospheric (RETRO) (http://retro.enes.org) (0.50 × 0.50) monthly 1960–2000 emission base and GOCART background emission data. The model uses anthropogenic emissions from a number of global and regional inventories, biomass burning emissions from the Global Fire Emission Database, and biogenic emissions from Model of Emissions of Gases and Aerosols from Nature (MEGAN) [23]. This data set is inserted in PREP-CHEM-SRC to estimate the emission fields over the user-specified simulation domain.

Additionally, a mobile emission inventory in the urban areas in Macedonia is added, with special emphasis on the city of Skopje. The mobile emission implemented in the system represents an emission inventory updated by the Ministry of Environmental and Physical Planning (MOEPP). The data emissions distributed by the GNFR sectors with a grid resolution of $0.1° × 0.1°$ lat/long are part of the Central Data Repository of European Environment Information (EIONET) and Observing Network of Long-Range Transport and Pollution Convention (CLRTAP). The inventory data set represents emissions in kt (kilotons) per year for each grid cell by chemical species (including CO and NOx). The emission rates and the coordinates are positioned in the central point of each given grid. The emissions are provided using the gridded mobile inventory data sourced by the MOEPP with a resolution of $0.1° × 0.1°$ lat/long with a surface area of about 11.10 km × 8.539 km or 94.78 km^2 approximately which corresponds at 40° latitude. The domain-averaged emission rates of CO and NOx for the central point of each grid box are then calculated by the given emission rates from the four adjacent points, using a bilinear interpolation method.

The system employed four single model configurations defined on a Lambert projection. The basic numerical integration is performed with 5-km horizontal grid resolution centered at 41.55° N, 21.45° E and covers North Macedonia, with parts of Serbia, Bulgaria, Albania, and Greece. The grid network contains 70 × 70 grid points in both the east-west and north-south directions. The vertical grid in the model is composed of 35 levels from the surface to about 30 km with 10 levels within 1 km above the model surface.

The model is initialized by the real boundary conditions using NCAR-NCEP's final analysis (FNL) data [24] having a spatial resolution of 0.25° × 0.25° (~27.7 km × 27.7 km) and a 6-h temporal resolution or NCEP GFS data with the same spatial resolution.

The described setup is what we will later refer to as the initial model and we will use this as the referent model that we will further improve [19]. Although it is a relatively good model for predicting the peaks of urban air pollution in winter, the model as described shows a temperature dependent bias (as shown in Fig. 4). The cause for this bias will be presented in the next section, following with our proposed improvements.

3 Analysis of Measured Pollution Data

In order to understand the distribution of the air pollution and find patterns of correlation, we analyzed the publicly available pollution measurements data. The sources we used are a combination from the pollution monitoring stations operated by the MOEPP and the ones from the various other networks (crowdsourced sensors, experimental and research sensors). The available data was measured over a time period of few years and it is publicly available [25].

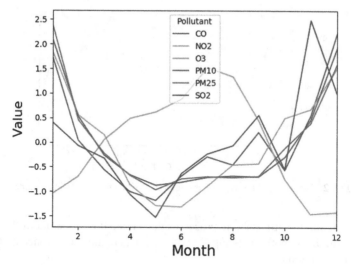

Fig. 1. Normalized monthly distribution of various pollutants

The data we were mostly interested in were the PM_{10} measurements. Most of the measurement stations have sensors that measure this parameter, and the fact that it is a pollutant that often gets the public attention due to the extreme measured values in winter, makes it the most interesting parameter to analyze and predict. The preprocessing of the data we did was minimal: removal of negative values and removal of values above $1300\,\mu g/m^3$ for the PM_{10} and $PM_{2.5}$. The upper limit was set at this value, as it is slightly above the official confirmed record high measured value, for which the initial model has been specifically evaluated for [9]. Therefore, the values above were considered outliers. We should mention here that we should be very careful with the filtering of outliers, as most of the standard outlier filters would likely fail by removing the rare extremely high values of air pollution, and might label them as outliers.

The rest of the pollutants were filtered with the threshold of 5 sigma values of the data. This is larger than the standard of 3 sigmas, due to the fact that the distribution here is not Gaussian: it is significantly skewed, with a cutoff below 0. Another reason we were not very interested in a more thorough outlier removal is that we only intend to use the data in an aggregated form, by taking monthly and hourly averages over the entire available data. The eventual outliers would not seriously affect the conclusions that we would draw from the aggregated data, therefore we consider that the effort to remove the outliers in this case would be unnecessary for our purpose.

The plots that are crucial in improving the model can be seen in Fig. 1 and Fig. 2. Figure 1 shows the average monthly distribution of various pollutants. As we can see, for almost all of the pollutants, the peak concentrations occur in winter, while the measured pollution in summer was relatively low. The only exception here are the tropospheric ozone concentrations, which show a peak in summer. This is related to the increased

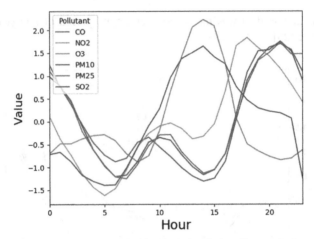

Fig. 2. Normalized hourly distributions of various pollutants in January

solar radiation, and the atmospheric processes that create the ozone out of the oxygen. We can confirm that hypothesis in Fig. 2 and Fig. 3, where the ozone concentrations are the highest around noon.

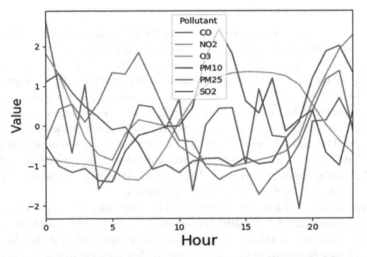

Fig. 3. Normalized hourly distributions of various pollutants in July

Figure 2 shows a bimodal distribution of the rest of the pollutants, except for SO2, for which we determined that the data quality is low and therefore we would not analyze it. The bi-modal distribution of PM10, PM2.5, CO and NO2 has two peaks: a lower one in the morning hours, corresponding with the early commute, and a higher one in the late afternoon, starting with the evening commute and continuing with the sharp increase, peaking before midnight. We conclude that the morning peak is caused by the traffic pollution and partly due to the home heating. The process of pollution caused by home heating is more pronounced in the late evening: at these hours, the traffic calms down from the evening commute, and most of the population is at home. Given that a majority of the population uses solid fuels for space heating [26], this is a plausible hypothesis [27].

Another hint to this hypothesis is the plot in Fig. 3. Here we present the data from July in a similar way as in Fig. 2. However, we do not see such a pronounced bimodal distributions with large peaks over night. All of the other pollution sources, like traffic and industry are present in July, however, the pollution from domestic heating is lacking. This supports the hypothesis that the combustion of solid fuels for space heating is a significant cause for the increased pollution at winter. In the next chapters, we will propose a modification of the setup for the WRF-Chem model presented in the previous chapter, and will experimentally test a model that takes modified initial chemical conditions (pollutants emissions) based on the temperature.

4   Temperature Dependent Model

All the data we have seen so far points towards the need of a temperature dependent model for the initial chemical conditions. Both the behavior of the measured air pollution, and the temperature-dependent bias of the model are strong indicators for that. We propose that by scaling the initial chemical conditions by the predicted temperatures over our simulation domain, we would expect improvements in the predictions. A concrete proposal for a temperature dependent model will be presented in this chapter.

If we take the average value of PM_{10} at each degree of Celsius predicted by our model and subtract the average measured value at the stations for the same temperature, we will get a measure that represents the average temperature dependent bias of the model. This is shown in Fig. 4, where we present the measured and simulated data for the first half of 2019 over all of the measurement points in the country (X- temperature in Celsius, Y- Bias). The simulated data is sourced by the initial mode, running online for day to day predictions [9]. We can notice that the initial setup of the model is highly temperature biased, namely, there is a large underestimation of the pollution at low temperatures. The temperature dependent bias seems to have a negative linear dependence by the temperature, but only at temperatures below 15 °C. In temperatures above that point, the model does not show a temperature dependent bias.

The authors of [9] describe that the user defined emissions were defined on a total yearly basis, and were fed to the model indiscriminately of the season. This deviates from what we have presented in Sect. 3, where a seasonal variation was shown. As we have explained, we expect that the seasonal variation is due to the hypothesis we postulated for domestic space heating emissions. Namely, these emissions occur almost exclusively in winter, and are expected to be temperature dependent. The negative temperature dependence may be explained by the fact that the heating intensity, and therefore the combustion of solid fuels is higher at lower temperatures.

In order to improve the temperature dependent bias of the model seen in Fig. 4, we will do a modification in the initial chemical conditions to include temperature dependent scaling. The scaling is done only in the data provided by the MOEPP for urban mobile emissions. More specifically, we only scale the data from the sector of "other stationary

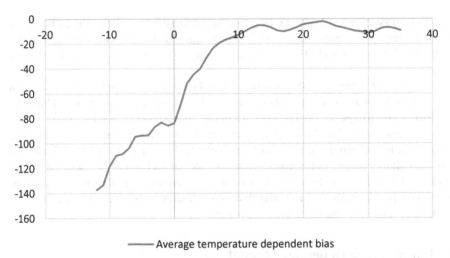

Fig. 4. Average temperature-dependent bias of the initial WRF-Chem setup over all measurement points

combined" emissions, which contains, and mostly consists of the emissions coming from domestic heating sources.

The scaling of the emissions is done in the following way: for each emission cell of the grid that contains an urban area, we multiply the "other stationary combined" source by a factor F. The factor depends on the average predicted temperature for that cell box for the period we intend to run the simulation. The average temperature for the cell box is calculated by averaging the predicted temperatures for all of the simulation nodes within that cell, over the entire period we intend to do our simulation run. The visualization of the cell boxes where the pollution is given, and our simulation nodes from which we get the temperatures are given in Fig. 5.

Once the average temperature is calculated, we need to calculate the scaling factor F. For average temperatures above 15 °C, $F = 0.1$, and for temperatures below that point, we increase F by 0.15 for the drop of each degree Celsius below 15 °C. The formula for scaling is the following one:

$$F = 0.1, \ T >= 15C; \tag{1}$$

$$F = 0.1 + (0.15 * (15C\text{-}T)), \ T < 15 \tag{2}$$

These modified user defined emissions are fed into the PREP-CHEM-SRC program, and the rest of the simulation workflow is unchanged from the one presented in Sect. 2. With this approach, we only change the emission that come from the domestic heating sources, by making them follow a temperature-dependent formula which is hypothesized to match the reality. The reasoning behind this is that the domestic heating is intensified when the temperatures drop, and is almost non-existent when the average temperatures are above 15 °C.

We should also point that our initial conditions are static, i.e. we do not modify these conditions during the simulation run. Feeding dynamic initial conditions into the

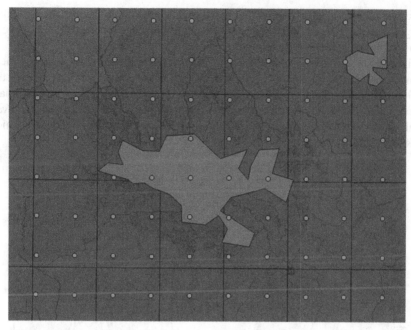

Fig. 5. A map of the wider Skopje area with the GIS data we use in our system

WRF-Chem model is a challenging task. However, when using the model for day-to-day predictions, we would rarely use simulation domain larger than 3 days, over which the average temperatures are very unlikely to change. Thus, we expect that the cost benefit of integrating over dynamic initial conditions would be small.

5   Evaluation

We have evaluated the proposed model by doing 24 simulation runs for intervals of five days each. The simulation periods were distributed evenly across the year, as we have simulated five day intervals starting at the first and 16-th day of every month. In this way, we expect to simulate a variety of different meteorological and meteo-chemical conditions, as well as be able to produce enough data to evaluate the temperature-related bias of the model for each of the measurement point locations separately.

Due to the regional variation and bias of the model, we selected the wider Skopje area for a detailed evaluation, given that the temperature dependent bias, and the general bias was calculated over this area. The model is easily extensible over the other regions as well in the same way as done for this region, but due to the high population density, we decided to focus our attention on the Skopje region. The methodology of evaluation was to calculate the predictions at every point in space where a measurement station exists. As the measurement stations do not match with the nodes in our simulation grid, the predictions at these specific points were calculated by taking the interpolated values of the nearby simulation nodes. We took all the values at the simulation nodes in radius of 5 km around the measurement point, and calculated the interpolated value by taking a weighted average, where the weights are inversely dependent on the distance between the measurement point and the simulation nodes.

Figure 6 shows the measured values for PM10 concentrations at the Karposh measurement station, for a period of 5 days during winter. We can notice that although the initial model predicts the pattern of the curve relatively well, it does tend to overestimate the pollution over the entire period. We can also notice that with our proposed improved model, we are able to follow the observed curve with our predictions more tightly, which points to a significant improvement.

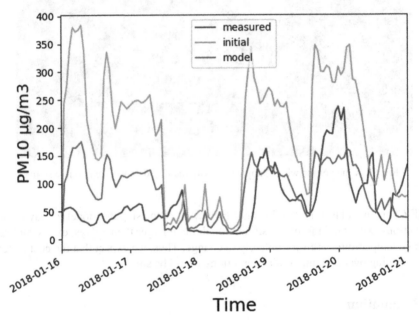

Fig. 6. Measured and predicted values for Karposh in winter

A similar pattern, though with a significantly larger differences can be observed for a period of 5 days during the summer. Due to the previously mentioned temperature dependent bias, the initial model tends to significantly overpredict the pollution levels in these conditions. On the other hand, our improved model tends to drastically minimize the discrepancy between the measured and predicted values during summer, when the pollution emissions are much lower than the yearly averages (Fig. 7).

It is worth noting that the improvements proposed in these paper do not change the shape of the curve, and we can still notice the exact same spikes and holes in the curve in both of the charts. This is due to the fact that our work was focused on modeling the initial and boundary chemical conditions in the simulation. These are kept constant over the entire simulation run, so it is reasonable to expect that they might only linearly shift the curve by a certain amplitude factor. In order to be able to change the curve shape, we would need to intervene in the chemical simulation model itself, or simulate with

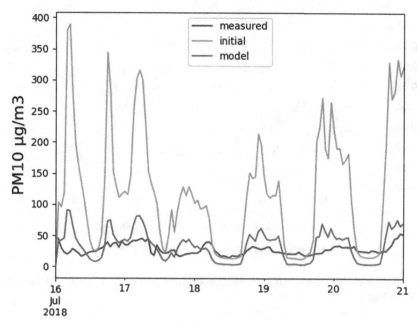

Fig. 7. Measured and predicted values for Karposh in summer.

dynamic initial and boundary conditions. This might be interesting for future work, but it is well beyond the scope of our current research.

Figure 8 (should put a table here instead) shows the bias of the model at two stations, Centar and Karposh. We can note that our model shows a smaller bias in general for both stations, as well as a flatter curve. The flatter curve is a clear indicator that in comparison to the base model, the temperature dependent bias has been decreased by our proposed improvements.

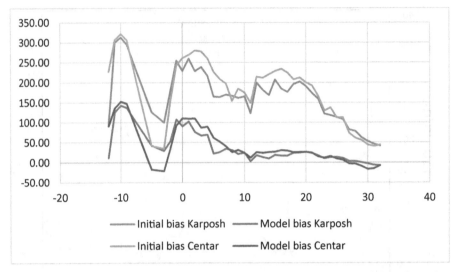

Fig. 8. Temperature dependent bias of the initial model versus our improved model, for the stations Centar and Karposh

6 Conclusion and Future Work

In this paper, we first presented a general overview of the air pollution prediction models, which are becoming increasingly important in simulating the air pollution in urban areas. Our detailed overview was focused on the WRF-Chem model, where we evaluated a concrete setup of the model, tailored for the conditions of the wider Skopje region. Then we were able to propose a possible improvement of the initial chemical conditions used by the model. Our hypothesis was supported by the data measured from the network of air pollution sensors, which also indicated to a temperature dependent bias of the model. Combining the two findings, we managed to quantitatively measure the deviations of the model from the observed values, and based on that we proposed an improvement for the initialization of the initial chemical conditions.

Our proposed change to the model resulted in measurable improvements in predictions. The total bias of the model, evaluated over the entire simulated data, was decreased from 205.28 to 40.15 for the measurement station in Centar, 179.67 to 32.79 for Karposh, and the charts presented show that we are now able to make predictions that tightly follow the curves of the observed values.

However, the new model still has regional variations, which points out to a regional-dependent bias. In this paper we focused our improvements for the wider Skopje area, but any future work should address this issue of regional bias, by providing specific scaling factors for every region separately, with the same methods elaborated here that we employed for the Skopje region. This issue of regional variations might also be caused by the relatively coarse simulation grid we use (5 km × 5 km horizontally). Due to the constraints on computational power, we were not able to test a model with a finer simulation mesh and this could also be an interesting topic for further research.

The coarse mesh might also be the cause for the regional variations, which is to be investigated as well.

Finally, our future work may also be directed towards a model that will have a learning property and be able to improve itself dynamically on the base of the calculated bias of its previous predictions. This would be useful in practical application, where the model would be brought to online service for making day to day pollution predictions.

References

1. MojVozduh homepage (2020). https://mojvozduh.eu/. Accessed 28 May 2020
2. SkopjePulse homepage (2020). https://skopje.pulse.eco/. Accessed 28 May 2020
3. European Commission environment agency, Air pollution limits page (2020). https://ec.eur opa.eu/environment/air/quality/standards.htm. Accessed 28 Mya 2020
4. Windy homepage (2020). https://windy.com. Accessed 23 May 2020
5. MOEPP of N. Macedonia, Air quality modeling page (2020). http://air.moepp.gov.mk/?page_i d=331. Accessed 18 May 2020
6. Sofiev, M., Siljamo, P., Valkama, I., Ilvonen, M., Kukkonen, J.: A dispersion modelling system SILAM and its evaluation against ETEX data. Atmos. Environ. **40**(4), 674–685 (2006). https:// doi.org/10.1016/j.atmosenv.2005.09.069
7. EURAD home page (2020). http://db.eurad.uni-koeln.de/en/forecast/eurad-im.php. Accessed 03 May 2020
8. Strunk, A., Ebel, A., Elbern, H., Friese, E., Goris, N., Nieradzik, L.P.: Four-dimensional variational assimilation of atmospheric chemical data – application to regional modelling of air quality. In: Lirkov, I., Margenov, S., Waśniewski, J. (eds.) LSSC 2009. LNCS, vol. 5910, pp. 214–222. Springer, Heidelberg (2010). https://doi.org/10.1007/978-3-642-12535-5_24
9. Spiridonov, V., Jakimovski, B., Spiridonova, I., Pereira, G.: Development of air quality fore-casting system in Macedonia, based on WRF-Chem model. Air Qual. Atmos. Health **12**(7), 825–836 (2019). https://doi.org/10.1007/s11869-019-00698-5
10. UCAR WRF homepage (2020). https://www.mmm.ucar.edu/weather-research-and-foreca sting-model. Accessed 20 May 2020
11. UCAR WRF-Chem homepage (2020). https://www2.acom.ucar.edu/wrf-chem. Accessed 20 May 2020
12. PREP-CHEM-SRC online user guide (2020). http://ftp.cptec.inpe.br/brams/BRAMS/docume ntation/guide-PREP-CHEM-SRC-1.5.pdf. Accessed 20 May 2020
13. EIONET homepage, data section for the Republic of North Macedonia (2020). https://cdr.eio net.europa.eu/mk. Accessed 28 May 2020
14. Ministry of Environment and Physical Planning of the Republic of North Macedonia: Infor-mative Inventory Report 1990–2016 Under the Convention on Long-Range Transboundary Air Pollution (CLRTAP) for North Macedonia (2016). http://airquality.moepp.gov.mk/airqua lity/wp-content/uploads/2012/05/IIR_Macedonia_2016.pdf. Accessed 30 May 2020
15. Skamarock, W.C., et al.: A description of the advanced research WRF version 3. In: NCAR Technical Notes, No. NCAR/TN475+STR, University Corporation for Atmospheric Research (2008). https://doi.org/10.5065/d68s4mvh
16. Grell, G.A., et al.: Fully-coupled online chemistry within the WRF model. Atmos. Environ. **39**, 6957–6975 (2005). https://doi.org/10.1016/j.atmosenv.2005.04.027
17. Frost, G.J., et al.: Effects of changing power plant NOx emissions on ozone in the eastern United States: proof of concept. J. Geophys. Res. **111**, D12306 (2006). https://doi.org/10. 1029/2005jd006354

18. Stockwell, W., Middleton, P., Chang, J.: The second generation regional acid deposition model chemical mechanism for regional air quality modeling. J. Geophys. Res. **951**, 16343–16367 (1990). https://doi.org/10.1029/jd095id10p16343

19. Ackermann, I.J., Hass, H., Memmsheimer, M., Ebel, A., Binkowski, F.S., Shankar, U.: Modal aerosol dynamics model for Europe: development and first applications. Atmos. Environ. **32**, 2981–2999 (1998). https://doi.org/10.1016/s1352-2310(98)00006-5

20. Schell, B., Ackermann, I.J., Hass, H., Binkowski, F.S., Ebel, A.: Modeling the formation of secondary organic aerosol within a comprehensive air quality model system. J. Geophys. Res. **106**(D22), 28275–28293 (2001). https://doi.org/10.1029/2001jd000384

21. Freitas, S.R., et al.: The Brazilian developments on the Regional Atmospheric Modeling System (BRAMS 5.2): an integrated environmental model tuned for tropical areas. Geosci. Model. Dev. **10**, 189–222 (2017). https://doi.org/10.5194/gmd-10-189-2017

22. Olivier, J.G.J., van Aardenne, J.A., Dentener, F.J., Pagliari, V., Ganzeveld, L.N., Peters, J.A.H.W.: Recent trends in global greenhouse gas emissions: regional trends 1970–2000 and spatial distribution of key sources in 2000. Environ. Sci. **2**(2–3), pp. 81–99 (2005). https://doi.org/10.1080/15693430500400345

23. Emmons, L.K., et al.: Description and evaluation of the Model for Ozone and Related chemical Tracers, version 4 (MOZART-4). Geosci. Model. Dev. **3**, 43–67 (2010). https://doi.org/10.5194/gmd3-43-2010

24. NCEP datasets site. https://rda.ucar.edu/datasets/ds083.2/. Accessed 23 Apr 2020

25. Data repository of MojVozduh (2020). https://github.com/jovanovski/MojVozduhExports. Accessed 21 Mar 2020

26. State institute of statistics of the Republic of North Macedonia, Statistical review Industry and Energy, Statistical review 6.4.15.03/836. http://www.stat.gov.mk/Publikacii/6.4.15.03.pdf

27. Mirakovski, D., et al.: Sources of urban air pollution in Macedonia – behind high pollution episodes. In: International Scientific Conference GREDIT 2018 – Green Development, Green Infrastructure, Green Technology, Skopje, 22–25 March 2018 (2018)

Smart City Air Pollution Monitoring and Prediction: A Case Study of Skopje

Jovan Kalajdjieski, Mladen Korunoski, Biljana Risteska Stojkoska,
and Kire Trivodaliev[✉]

Faculty of Computer Science and Engineering, Ss. Cyril and Methodius University,
1000 Skopje, Macedonia
{jovan.kalajdzhieski,biljana.stojkoska,kire.trivodaliev}@finki.ukim.mk,
korunoski.mladen@students.finki.ukim.mk

Abstract. One of the key aspects of smart cities is the enhancement
of awareness of the key stakeholders as well as the general population
regarding air pollution. Citizens often remain unaware of the pollution in
their immediate surrounding which usually has strong correlation with
the local environment and micro-climate. This paper presents an Internet of Things based system for real-time monitoring and prediction of
air pollution. First, a general layered management model for an Internet of Things based holistic framework is given by defining its integral
levels and their main tasks as observed in state-of-the-art solutions. The
value of data is increased by developing a suitable data processing sub-
system. Using deep learning techniques, it provides predictions for future
pollution levels as well as times to reaching alarming thresholds. The
sub-system is built and tested on data for the city of Skopje. Although
the data resolution used in the experiments is low, the results are very
promising. The integration of this module with an Internet of Things
infrastructure for sensing the air pollution will significantly improve overall performance due to the intrinsic nature of the techniques employed.

Keywords: Internet of Things · Smart city · Air pollution
monitoring · Air pollution prediction

1 Introduction

One of the key problems of major urban areas in developing and industrial countries is air pollution, especially when measures for air quality are not available, or
are minimally implemented or enforced [1]. According to the report of the World
Health Organization (WHO) [2], around 91% of the world's population lives in
places where air quality exceeds the WHO guidelines, and around 4.2 million
deaths every year can be directly linked to exposure to outdoor air pollution.
Chronic exposure to air pollution increases the risk of cardiovascular and respiratory mortality and morbidity, while acute short-term inhalation of pollutants
can induce changes in lung function and the cardiovascular system exacerbating

V. Dimitrova and I. Dimitrovski (Eds.): ICT Innovations 2020, CCIS 1316, pp. 15–27, 2020.
https://doi.org/10.1007/978-3-030-62098-1_2

existing conditions such as ischemic heart disease [3,4]. In less developed countries, 98% of children under five are exposed to toxic air. This makes air pollution the main cause of death for children under the age of 15, killing 600,000 every year [5]. The World Bank estimates \$5 trillion in welfare losses worldwide due to air pollution premature deaths [6]. Air pollution also contributes to climate changes which increases premature human mortality [7].

Urban outdoor air pollution, especially particulate matter, remains a major environmental health problem in Skopje, the capital of North Macedonia. Long-term exposure to PM2.5 caused an estimated 1199 premature deaths and the social cost of the predicted premature mortality in 2012 due to air pollution was estimated at between 570 and 1470 million euros according to Martinez et al. in [8]. Additionally, in the same year, there have been 547 hospital admissions from cardiovascular diseases, and 937 admissions for respiratory disease due to air pollution. The study also infers that if PM2.5 was reduced to EU standards (25 $\mu g/m^3$ at that timepoint), it could have averted an estimated 45% of PM-attributable mortality, but if PM2.5 was reduced to the WHO Air Quality Guidelines (10 $\mu g/m^3$ at that timepoint), around 77% of PM-attributable mortality could have been averted, which could have provided a substantial health and economic gain for the city. A more recent report on the Air quality in Europe [9], done by the European Environment Agency (EEA), shows that the situation in North Macedonia has deteriorated even further. It shows that the PM2.5 average exposure indicator for the period of 2015–2017 based on measurements in urban and suburban stations is 51 $\mu g/m^3$, which is substantially over the EU standard of 20 $\mu g/m^3$.

Many of the world governments deploy and operate stations for air quality monitoring and make the acquired data publicly available. These stations have high quality sensors which allow to sense a wide range of pollutants (like CO, NO2, SO2, O3, PM - particulate matter, etc.). However, the high costs of installing and maintaining them limits their number. In such cases, the low spatial resolution is resolved by using mathematical models that estimate the concentrations of the pollutants over the complete geographical space of interest. Although these models are complex and incorporate various input parameters such as meteorological variables, they can still be inaccurate (due to highly variable meteorological conditions [10]) which can lead to unsubstantiated inferences [11]. The Ministry of environment and physical planning of Republic of North Macedonia has build an infrastructure of 21 stations. 2 of these stations are mobile and the others are distributed throughout the country in 3 major regions: Skopje region (7 stations), Western region (6 stations) and Eastern region (6 stations). These stations apart from the extremely low spatial resolution, have problems reporting data due to sensor malfunctioning for longer periods of time, poor interconnectedness with the central reporting site, and even further not all of them have the appropriate sensors for measuring all relevant pollutants. Therefore, building a proper monitoring infrastructure is the first step towards healthier air.

Once quality data are available it can be used to extract deeper knowledge for pollution. Building air pollution prediction systems allows for the prediction of the air quality index (AQI), the value of each pollutant (i.e. PM2.5, PM10, CO2 and etc.) and high pollution areas. With such systems available, governments can employ smarter solutions for tackling the problem preemptively. To date, there have been many proposed solutions for predicting air pollution, but generally, these models can be classified into two types. The first type generate models that track the generation, dispersion and transmission process of pollutants. The predictive results of these models are given by numerical simulations. On the other hand, the second type of models are statistical learning models or machine learning models. These models attempt to find patterns directly from the input data [12].

The aim of this research is to propose an Internet of Things based system for real-time monitoring and prediction of air pollution. The first objective is to define a general layered management model for an IoT based holistic framework by defining its integral levels and their main tasks as observed in state-of-the-art solutions. Increasing the value of data by developing a suitable data processing sub-system is the second objective. Employing advanced machine learning techniques, especially deep learning, should increase the robustness of the system and provide insight on future trends which is essential for implementing appropriate policies. The final objective is to provide a proof-of-concept for the system considering a case-study for the city of Skopje.

The rest of this paper is organized as follows. State-of-the-art IoT architectures and intelligent data processing in air pollution is covered in the next section. In the third section the proposed IoT based system is presented. The case-study for the city of Skopje is given in the fourth section. Finally, this paper is concluded in the fifth section.

2   Related Work

A cloud-based architecture containing multiple data collection nodes was designed by Sendra et al. in [13]. The nodes are either mobile or static and store the data collected in local databases, but one centralized database is used to integrate the data collected from all the sensors. The integration steps also include the user's opinion left from their mobile device. The main purpose is to have a collaborative decision and alerting system. A network of air sensors connected on Arduino chips is designed in [14]. The pollution measurements from the sensors are sent on a cloud platform and the data is used in a mobile application. The application contains a map where a user chooses two points and the application shows the pollution between the two points. Another approach composed of Arduino chips connected with sensors is proposed in [15]. However, in this approach the sensors are static and are placed on a college campus. The data is stored on a centralized computer and later it can be used for visualizations.

An architecture consisting of multiple static air pollution sensors connected to a Raspberry Pi controller is proposed in [16]. The pollution measurements are collected and sent out to a cloud platform to be stored. The main goal is having a monitoring system which alerts the users when the pollution values are higher than the predefined threshold. Another approach using a Raspberry Pi controller is proposed in [17]. The architecture employs mobile devices to measure the noise pollution using mobile devices. The audio recordings along with GPS coordinates and metadata are sent out and stored in a MongoDB database. The data can then be used for visualisations of the pollution.

A cloud architecture employing the master/slave communication model is proposed by Saha et al. in [18]. Air, water and noise data is collected by sensors and sent out to the cloud for the purpose of monitoring pollution. A noise pollution monitoring network architecture consisting of three tiers is designed in [19]. The bottom tier consists of mobile and static sensors. The middle tier is built of relay nodes which collect the data from the sensors and deliver it to the gateways in the top layer. The gateways deliver the data to the cloud system. A slightly different approach is proposed by Zhang et al. in [20]. They create a knowledge graph which fuses data from social media data, air sensors data, taxi trajectory and traffic condition. The data is firstly converted to abstract entities with semantic data included. The city in the knowledge graph is divided into blocks and external databases block knowledge is collected. The main purpose of the knowledge graph is to detect and predict pollution with semantic explanation of the obtained results. It can also be used to analyse traffic patterns. Ahglren et al. in [21] proposed an architecture containing multiple sensors which use a publish/subscribe protocol and Message Queuing Telemetry Transport (MQTT) to communicate in an open format with the gateways. The gateways send out the data to a cloud architecture to be further preprocessed before it can be requested. The data can be requested as raw or aggregated by predefined time frames (hourly, daily, weekly or monthly averages of the values). A cloud based architecture is also implemented in [22]. The architecture consists of multiple static sensors placed on streetlight poles. The pollution measurements are sent out to a sink node which is their single point of contact. The sink nodes deliver the data to the cloud service to be stored in a database. The cloud can then deliver fine-grained measurements or average values of pollution in a time frame of 7 days.

Apart from many IoT proposed architectures for tackling the problem of monitoring pollution and implementing alerting logic, there are also many approaches tackling the problem of predicting pollution and pollution areas. One such approach is used in [23], where the authors use the CityPulse open dataset. The dataset consists of air data collected by 449 sensors placed besides traffic lights in the city of Brasov - Romania. The main focus of this approach is finding low and high pollution areas by analyzing the density of the ozone using K-means clustering. This approach although simpler than many others, still provides meaningful results.

However, in recent years, deep learning has been the main technique for air pollution prediction. While there have been many different models proposed, they mainly use the same pollution data. In all of the models, pollution measurements at specific time and location are taken into consideration. The pollutants usually measured are the particulate matter (i.e. PM2.5 and PM10) and gaseous species (i.e. NO2, CO, O3 and SO2). The data in the different approaches has different time intervals, but even so the main logic is the same. They also integrate meteorology data such as such as humidity, temperature, wind speed and rainfall. There are some that even consider weather forecast data [24,25]. The preprocessing is mainly the feature extraction part and it consists of principal component analysis, cluster analysis, factor analysis and discriminant analysis. Not much information has been provided on fusing the data from multiple sources, but where it is provided, we can conclude that only simple techniques such as matching by time is used. Most of the approaches execute the preprocessing stage separate from the model [12,25–28], but there are some approaches that incorporate this stage into combining the feature analysis and interpolation directly into the model [24,29]. When talking about the models, they can be mainly separated into two different categories: models predicting pollution level [24,25,28,29], (such as air quality index or air pollution index) and models predicting the level of pollutants (such as PM2.5, PM10, NO2 etc.) [12,26,27,30]. For the first types of models the first task is to label the fine-grained data so it can be used as an input to the model. On the other hand, the second type predicts the actual values for the pollutants, so no labeling is needed. The models used in both types are mainly neural networks [24,29,30], recurrent neural networks [12] (RNN) and LSTM (long short-term memory) networks as special types of RNN [28]. There are some approaches that use autoencoder model [26], merge neural networks with predictors for example linear predictors [25], Bayesian networks and multilabel classifiers [27]. After the models have been trained, different methods for evaluation are used, but mainly consist of: root mean square error, mean absolute error, mean absolute percentage error, mean prediction error, relative prediction error.

The Weibul-time-to-event Recurrent Neural Network (WTTE-RNN) [31] is a relatively new model for time-to-event prediction, but it has been successfully used in medicine [32], predictive network diagnostics [33], as well as in state-of-the-art video processing and predictive modelling [34].

3   Internet of Things Based System for Air Pollution Monitoring and Prediction

Air quality monitoring and control is an essential part of the concept of smart city which is becoming the standard to which both developing and developed countries aspire, thus the public mindfulness for the process is high. In this section we are going to explain in detail our framework for air pollution monitoring and prediction based on Internet of Things as depicted in Fig. 1 for which the general layered management model is depicted in Fig. 1. The main tasks that should be performed at each level are described in the following subsections.

3.1 Sensing Layer

Sensors are used to sense, actuate, process data and communicate. To successfully sense and actuate, A/D and D/A conversion is needed. The sensors sense and send data periodically, wirelessly or wired, to the gateway. Sensed data can also be sent directly to the cloud, if protocols allow it. Both static and mobile sensors that measure the level of different pollutants need to be considered in order to achieve good spatial data resolution. If possible, sensor should perform basic data processing before sending out the data. Mutual validation of the sensed data can be done by different types of sensors when they spatially overlap (edge computing).

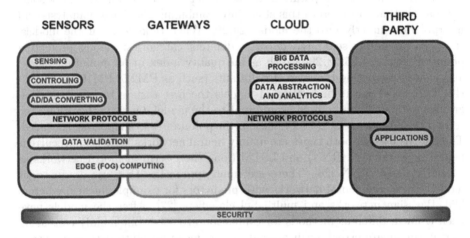

Fig. 1. A general layered management model

3.2 Communication Gateways Layer

For collecting raw and/or processed data from the sensors, gateways are employed. The gateways forward the data to the cloud. In order to reduce the data flow towards the cloud, whenever possible, the gateways should perform local data processing (fog computing). Edge/fog computing is very important in terms of the robustness of the infrastructure. Lightweight local processing algorithms can reduce the need for transmission, thus saving energy and avoiding latency issues and saturation of the communication channels.

The gateways can also act as a local scheduler, load balancer or regulator, sending out commands to the sensors. Furthermore, because the devices usually cannot communicate with each other, gateways provide interoperability between them. The communication between the sensing layer and the cloud should be effective, robust and operationally consistent so various possibilities should be considered (e.g. mobile network, LoRa, etc.).

3.3   Cloud Processing Layer

The most complex part of our air pollution monitoring and prediction system is the cloud, providing abstraction, data storage and analytics. Because of the high data volume, traditional approaches should be modified to meet the new requirements. New methods and algorithms based on machine learning techniques, time series processing and advanced analytics are to be employed. The data processing sub-system architecture is given in Fig. 2.

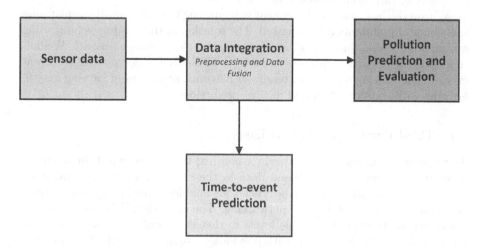

Fig. 2. Data processing sub-system architecture

Data Processing Sub-system. We have developed a more robust approach, taking into consideration the disadvantages of the previous models. The data preprocessing and integration module is the core feeding its results to the other modules for: pollution prediction and time-to-event prediction.

In the first step of data integration and preprocessing, we focus on integrating the data from various sources. Aside from pollution data, meteorological data are also needed, and they can be used from any available web service that provides such data (if our sensing layer does not provide it). The initial preprocessing of the data includes removing outliers and smoothing. Very often sensors encounter problems, data will be missing for the time-frame when a sensor was not working. This problem will be addressed using Deep Belief Networks (DBN) for generating missing data. The next step is to spatially discretize the data into regions using Delaunay triangulation. Finally, additional features from the time-series sensing data are extracted using Principal Component Analysis (PCA).

In the module for pollution prediction different deep learning models are built to predict the pollutant levels and the overall air quality as expressed by the AQI. For each spatial region, a corresponding Convolutional Neural Network

(CNN) and Recurrent Neural Network (RNN) is built. Different architectures for building the neural networks are evaluated as well as the possibilities for their combination to improve overall performance. The evaluation is performed using k-fold cross-validation for the initial building of the model. Once a model is deployed its predictive power is evaluated on the fly. This means that after we conduct a prediction, we would be checking this data with the real data collected to be able to create a back propagation and ensure the parameters of our models are correctly configured. Future developments will evaluate the feasibility of employing reinforcement learning.

Within the last module the system is predicting the hours until the alarming thresholds of pollutants are surpassed. The solution in this module is built using a framework based on survival analysis where a deep learning model, Weibul-time-to-event Recurrent Neural Network (WTTE-RNN), incorporates recurrent events, time varying covariates, temporal patterns, sequences of varying length, learning with censored data and flexible predictions.

3.4   Third-Party Applications Layer

Third-party applications use data on a non-real-time basis, which imposes the need to transform the event-based data in the cloud to a format suitable for query-based processing. This is a crucial for enabling third-party applications in a system with real-time IoT networking. The data should be stored persistently and abstracted at multiple levels so that they could be easily combined, recomputed and/or aggregated with previously stored data, with the possibility of some data coming from non-IoT sources. Even more importantly the different levels of abstraction will simplify the application access and usage since data will be presented in a manner required by applications [35].

4   Case-Study for the City of Skopje

As a proof-of-concept for the proposed data processing sub-system a case-study for the pollution data for the city of Skopje was considered. There were only 18 monitoring sites throughout the country with only 7 in Skopje, but not all sites measure the concentration of all the pollutants. Even though some sites are couple of hundred kilometers apart, all of them had to be considered when developing the models.

For the pollution prediction model, meteorological data obtained from the DarkSky API[1] are fused with the sensor data. The first step of data preprocessing was outlier analysis using the DBSCAN (Density-Based Spatial Clustering of Applications with Noise) method, which is an alternative of K-means clustering. The advantage of this method is that it automatically detects the number of clusters while maximizing the similarities between entries in the cluster. Next, data smoothing techniques were employed, which allows for efficient training of

---

[1] https://darksky.net/dev(lastaccessed30.04.2020).

the model, unbiased of the range of the features in the data. Seasonal exponential smoothing was used, because it allows to exponentially assign more weight to recent data points than to older data points, while considering the seasonality of the data. The next step was handling the missing records in the time-series per sensor, which occurs because of sensor malfunction or network errors. For this step, a Deep Belief Network was used consisting of many Restricted Boltzmann Machine (RBM) layers. This step augments the model by generating missing data points of interest, instead of removing them. A very important part of the model is the removal of the spatial relationship between the data, allowing us to view the data as time-series. For this step the Delaunay triangulation was used. While the Delaunay triangulation is a very powerful technique, it generated some regions which did not contain sensor measurements. To tackle this problem, empty regions were merged with the region that had the closest sensor to the empty region. After the data was split per region, the sensor and meteorological data was fused so we could obtain fine-grained data per sensor. Finally, Principal Component Analysis was conducted to further augment the data before training the model. For that purpose, different values for the number of principal components generated were tested. We found that 12 principal components performed best for our data.

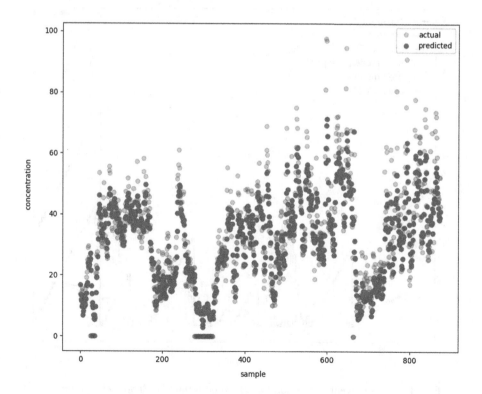

Fig. 3. PM10 pollutant concentration prediction performance for Skopje city center.

The final step of the pollution prediction model was building several Bidirectional Long-Short Term Memory (BiLSTM) models for the PM10 pollutant. The Bidirectional LSTM was used to learn the input sequence both forward and backwards and concatenate both interpretations. The ReLU activation function was used and the backpropagation was done using mean squared error (MSE). After fitting the model, the predicted and actual values were plotted and then the root mean squared error was calculated using the test set for PM10 pollutant for the monitoring site in the Skopje city center as shown in Fig. 3.

Regarding time-to-event prediction, the data is modified in a way that complies with the proposed WTTE-RNN objective function and the defined network architecture. A sliding window is used and the following steps are made:

- data for the past several days from the current time is captured, adding empty rows if necessary
- the time until the alarming threshold are surpassed is determined, for every row in the sample and whether that data is censored or not
- the data is split into train and test set and modified accordingly and the GRU model is fitted

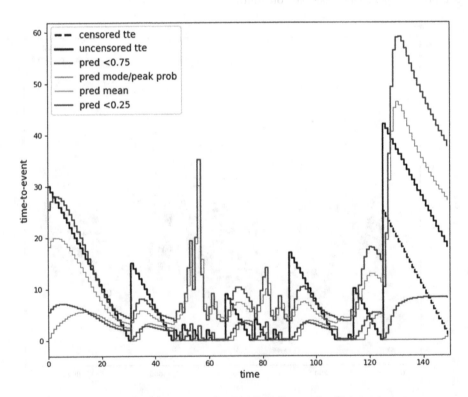

Fig. 4. Time-to-event prediction for PM10 pollutant for Skopje city center

In Fig. 4 The Weibull 0.25 and 0.75 quantiles are shown, mode and mean for the α and β parameters learned for PM10 pollutant measured in the Skopje city center. Although, they do not follow the data perfectly, we can observe the matching trends.

The mean squared error for all measurements in the Skopje city center is given in Table 1. The reason why the O3 and PM2.5 pollutants have larger values is because O3 has a lot of missing values that were generated in the augmentation step, while PM2.5 was measured by only three monitoring sites.

Table 1. Mean squared error for different pollutants

Pollutant	CO	NO2	O3	PM10	PM2.5	SO2
MSE	0.21	0.06	26.13	0.22	7.32	0.08

Considering the data resolution this simple prototype system works with, the results are very promising. Once the system is fully developed and the deployed IoT architecture feeds forward the data with significantly improved resolution, we believe that the performance will increase to scale.

5 Conclusion

This paper presents an Internet of Things based system for air pollution monitoring and prediction, composed of four layers cooperating together to enable its functionalities. This integral IoT framework is specific to the air pollution application domain, with the cloud being the central element in the system that serves not only to collect and store data, but also as a core data processing unit and a gateway to third-parties interested in developing applications. The operation of such environment is defined via a model with a set of specific tasks performed at each level to meet the system requirements.

The key element of any air pollution system in the context of smart city is the ability to extract deeper knowledge for the process by building real-time and future predictions. The data processing module proposed in this research has deep learning techniques at its core and provides increased robustness and reliability of the produced results. A case study for the city of Skopje showcases the performance of this module. Although the data resolution used in the experiments is low, the results are very promising. The integration of this module with an Internet of Things infrastructure for sensing the air pollution will significantly improve overall performance due to the intrinsic nature of the techniques employed. Furthermore, the general approach used in the presented system make it applicable in many domains for environment monitoring in smart cities.

Acknowledgements. This work was partially financed by the Faculty of Computer Science and Engineering, University Ss. Cyril and Methodius, Skopje.

References

1. World Health Organization: Ambient (outdoor) air quality and health. In: World Health Organization (2016)
2. WHO: Air pollution (2018). https://www.who.int/airpollution/en/
3. Brook, R.D., et al.: Particulate matter air pollution and cardiovascular disease: an update to the scientific statement from the American heart association. Circulation **121**(21), 2331–2378 (2010)
4. Xing, Y.F., Xu, Y.H., Shi, M.H., Lian, Y.X.: The impact of PM2. 5 on the human respiratory system. J. Thorac. Dis. **8**(1), E69 (2016)
5. WHO: More than 90% of the world's children breathe toxic air every day (2018). https://www.who.int/news-room/detail/29-10-2018-more-than-90-of-the-world%E2%80%99s-children-breathe-toxic-air-every-day
6. World Bank: Air pollution deaths cost global economy us$225 billion (2016). https://www.worldbank.org/en/news/press-release/2016/09/08/air-pollution-deaths-cost-global-economy-225-billion
7. Silva, R.A., et al.: Future global mortality from changes in air pollution attributable to climate change. Nature climate change **7**(9), 647–651 (2017)
8. Martinez, G.S., Spadaro, J.V., Chapizanis, D., Kendrovski, V., Kochubovski, M., Mudu, P.: Health impacts and economic costs of air pollution in the metropolitan area of Skopje. Int. J. Environ. Res. Public Health **15**(4), 626 (2018)
9. EAA: Air quality Europe - 2019 report (2019). https://www.eea.europa.eu/publications/air-quality-in-europe-2019
10. Yeganeh, B., Hewson, M.G., Clifford, S., Tavassoli, A., Knibbs, L.D., Morawska, L.: Estimating the spatiotemporal variation of no2 concentration using an adaptive neuro-fuzzy inference system. Environ. Model. Softw. **100**, 222–235 (2018)
11. Beelen, R., et al.: Effects of long-term exposure to air pollution on natural-cause mortality: an analysis of 22 European cohorts within the multicentre escape project. Lancet **383**(9919), 785–795 (2014)
12. Fan, J., Li, Q., Hou, J., Feng, X., Karimian, H., Lin, S.: A spatiotemporal prediction framework for air pollution based on deep RNN. ISPRS Ann. Photogramm. Remote. Sens. Spat. Inf. Sci. **4**, 15 (2017)
13. Sendra, S., Garcia-Navas, J.L., Romero-Diaz, P., Lloret, J.: Collaborative lora-based sensor network for pollution monitoring in smart cities. In: 2019 Fourth International Conference on Fog and Mobile Edge Computing (FMEC), pp. 318–323. IEEE (2019)
14. Dhingra, S., Madda, R.B., Gandomi, A.H., Patan, R., Daneshmand, M.: Internet of things mobile-air pollution monitoring system (IoT-Mobair). IEEE Internet Things J. **6**(3), 5577–5584 (2019)
15. Ali, H., Soe, J., Weller, S.R.: A real-time ambient air quality monitoring wireless sensor network for schools in smart cities. In: 2015 IEEE First International Smart Cities Conference (ISC2), pp. 1–6. IEEE (2015)
16. Kiruthika, R., Umamakeswari, A.: Low cost pollution control and air quality monitoring system using Raspberry Pi for Internet of Things. In: 2017 International Conference on Energy, Communication, Data Analytics and Soft Computing (ICECDS), pp. 2319–2326. IEEE (2017)
17. Jezdović, I., Nedeljković, N., Živojinović, L., Radenković, B., Labus, A.: Smart city: a system for measuring noise pollution. Smart Cities Reg. Dev. (SCRD) J. **2**(1), 79–85 (2018)

18. Saha, H.N., et al.: Pollution control using internet of things (IoT). In: 2017 8th Annual Industrial Automation and Electromechanical Engineering Conference (IEMECON), pp. 65–68. IEEE (2017)
19. Jin, J., Gubbi, J., Marusic, S., Palaniswami, M.: An information framework for creating a smart city through internet of things. IEEE Internet Things J. **1**(2), 112–121 (2014)
20. Zhang, N., Chen, H., Chen, X., Chen, J.: Semantic framework of internet of things for smart cities: case studies. Sensors **16**(9), 1501 (2016)
21. Ahlgren, B., Hidell, M., Ngai, E.C.H.: Internet of things for smart cities: interoperability and open data. IEEE Internet Comput. **20**(6), 52–56 (2016)
22. Zanella, A., Bui, N., Castellani, A., Vangelista, L., Zorzi, M.: Internet of things for smart cities. IEEE Internet Things J. **1**(1), 22–32 (2014)
23. Zaree, T., Honarvar, A.R.: Improvement of air pollution prediction in a smart city and its correlation with weather conditions using metrological big data. Turk. J. Electr. Eng. Comput. Sci. **26**(3), 1302–1313 (2018)
24. Yi, X., Zhang, J., Wang, Z., Li, T., Zheng, Y.: Deep distributed fusion network for air quality prediction. In: Proceedings of the 24th ACM SIGKDD International Conference on Knowledge Discovery & Data Mining, pp. 965–973 (2018)
25. Zheng, Y., Yi, X., Li, M., Li, R., Shan, Z., Chang, E., Li, T.: Forecasting fine-grained air quality based on big data. In: Proceedings of the 21th ACM SIGKDD International Conference on Knowledge Discovery and Data Mining, pp. 2267–2276 (2015)
26. Li, X., Peng, L., Hu, Y., Shao, J., Chi, T.: Deep learning architecture for air quality predictions. Environ. Sci. Pollut. Res. **23**(22), 22408–22417 (2016)
27. Corani, G., Scanagatta, M.: Air pollution prediction via multi-label classification. Environ. Model. Softw. **80**, 259–264 (2016)
28. Kök, İ., Şimşek, M.U., Özdemir, S.: A deep learning model for air quality prediction in smart cities. In: 2017 IEEE International Conference on Big Data (Big Data), pp. 1983–1990. IEEE (2017)
29. Qi, Z., Wang, T., Song, G., Hu, W., Li, X., Zhang, Z.: Deep air learning: interpolation, prediction, and feature analysis of fine-grained air quality. IEEE Trans. Knowl. Data Eng. **30**(12), 2285–2297 (2018)
30. Li, T., Shen, H., Yuan, Q., Zhang, X., Zhang, L.: Estimating ground-level pm2. 5 by fusing satellite and station observations: a geo-intelligent deep learning approach. Geophys. Res. Lett. **44**(23), 11–985 (2017)
31. Martinsson, E.: WTTE-RNN: Weibull time to event recurrent neural network. Ph.D. thesis, Chalmers University Of Technology (2016)
32. Gensheimer, M.F., Narasimhan, B.: A scalable discrete-time survival model for neural networks. PeerJ **7**, e6257 (2019)
33. Aggarwal, K., Atan, O., Farahat, A.K., Zhang, C., Ristovski, K., Gupta, C.: Two birds with one network: unifying failure event prediction and time-to-failure modeling. In: 2018 IEEE International Conference on Big Data (Big Data), pp. 1308–1317. IEEE (2018)
34. Neumann, L., Zisserman, A., Vedaldi, A.: Future event prediction: if and when. In: Proceedings of the IEEE Conference on Computer Vision and Pattern Recognition Workshops (2019)
35. Stojkoska, B.L.R., Trivodaliev, K.V.: A review of Internet of Things for smart home: challenges and solutions. J. Clean. Prod. **140**, 1454–1464 (2017)

Cloud Based Personal Health Records Data Exchange in the Age of IoT: The Cross4all Project

Savoska Snezana[1]([envelope]) [ID], Vassilis Kilintzis[2] [ID], Boro Jakimovski[3] [ID], Ilija Jolevski[1] [ID],
Nikolaos Beredimas[2], Alexandros Mourouzis[2], Ivan Corbev[3] [ID],
Ioanna Chouvarda[2] [ID], Nicos Maglaveras[2] [ID], and Vladimir Trajkovik[3] [ID]

[1] Faculty of Information and Communication Technologies, "St. Kliment Ohridski" University
– Bitola, Bitola, Republic of North Macedonia
{snezana.savoska,ilija.jolevski}@uklo.edu.mk
[2] Lab of Computing, Medical Informatics and Biomedical Imaging Technologies, Aristotle
University, Thessaloniki, Greece
billyk@med.auth.gr, {beredim,ioannach,nicmag}@auth.gr,
mourouzi@hotmail.com
[3] Faculty of Computer Science and Engineering, Ss. Cyril and Methodius University in Skopje,
Skopje, Republic of North Macedonia
{boro.jakimovski,ivan.chorbev,trvlado}@finki.ukim.mk

Abstract. The paper presents some of the results of the Cross4all project. The main aim of this paper is to highlight the proposed model of integrated cloud based cross border healthcare systems that introduce a PHR concept and support effort of introducing an e-health strategy where the patient is owner of data and the key point of data collection, using different manners of data acquiring, sometimes not connected with hospitals and country of living. The increase of e-health and health digital literacy in the region is also the point of interest needed for the acceptance of the concept and proposed model.

Keywords: E-services · Wearables · Personal health record (PHR) · Electronic health record (EHR) · Electronic medical record (EMR) · Electronic patient record (EPR) · Cloud computing

1 Introduction

The development of health information technologies in the last decade creates a broad range of new opportunities to improve the access to health services for citizens and especially healthcare delivery [1–3]. The typical examples of such trends are enabled by giving the patients access to their own health and treatment-related information [4–7].

The other important trend to consider is continual rise of health care costs [8]. An important segment for patients, physicians, and healthcare policy makers is whether it is possible to control costs while maintaining the quality of health care services on the same or higher level.

© Springer Nature Switzerland AG 2020
V. Dimitrova and I. Dimitrovski (Eds.): ICT Innovations 2020, CCIS 1316, pp. 28–41, 2020.
https://doi.org/10.1007/978-3-030-62098-1_3

Implementation of electronic health records (EHR), electronic medical records (EMR) and personal health records (PHR), is viewed as a critical step towards achieving improvements in the quality and efficiency of the health care system in many European countries [9]. EHR is defined as: 'a repository of information regarding the health status of a subject of care in computer process able form, stored and transmitted securely, and accessible by multiple authorized users' [10]. EHR can be understood as a repository of patient data in digital form, which stored and exchanged securely. EMR systems manage EHR, so they can be created, gathered, and managed by authorized healthcare providers within healthcare organizations. Data entered into the computer (e.g. X-ray, pathology and pharmacy data) which can be integrated into the record are called electronic medical record. EMR may be characterized as a partial health record under the hat of a health-care provider that holds a portion of the relevant health information about a person over their lifetime. An EPR is a sub-type of an EHR, used in a specific hospital or healthcare organization. According to the ISO/DTR 20514:2005 standard [10] it's defined as 'a repository of patient data in digital form, stored and exchanged securely and accessible by multiple authorized users. EHR differs from EMR and PHR on the basis of the completeness of the information the record contains and the designated custodian of the information. A PHR is often described as a complete or partial health record under the custodian of a person(s) that holds all or a portion of the relevant health information about that person over their lifetime.

EHR, EPR, EMR, Healthcare Information Systems (HIS) and the exchange of healthcare data are essential components of the IT infrastructure in healthcare.

Interoperability in healthcare refers to the possibility of exchanging health data between two or more interconnected systems. The data exchange should be interpreted and understood in the same way for all the interconnected systems. To reduce costs and improve quality of healthcare, collaboration of healthcare units from different locations is required. According to Metzger et al. [11], the interoperability of health information system is defined as: 'the capability of heterogeneous systems to interchange data in a way that the data from one system can be recognized, interpreted, used and processed by other systems'.

Many countries have implemented a network of interoperable EHR. The vision of these systems is to provide a secure and private EHR, record of health and healthcare history, data protection within the healthcare system. These data are available to authorized healthcare providers, medical staff and individuals-owners of the data.

The Cloud Computing (CC) technology offers eHealth systems the opportunity to enhance the features and functionality that they offer. CC can be used as platform to implement EHR systems because it offers great potential for quick access to healthcare information and other information concerning health.

In the case of deploying an EHR in CC environment system, the main advantage is the ability to share patient (health) records with other clinical and healthcare centers, and the integration of all the EHRs of a group of clinical centers. This aims to help to the medical staff to perform their everyday tasks [12, 13].

However, before applying cloud based EHR systems, issues related to data security, access control, patient privacy, and overall performance of the whole system must be addressed [14].

Rodrigues et al. [15] elaborate the issues and requirements for maintaining the security and privacy of EHR systems in CC environments and explain the requirements that a Cloud-based EHR management system must guarantee in terms of security. Some of the security issues that should be considered are role-based access, network security mechanisms, data encryption, digital signatures, and access monitoring.

This paper describes a cloud-based model of EHR for healthcare organizations, adapted to the needs of the distributed healthcare information systems. In the next section the concept of healthcare data exchange is highlighted and supported with explanation of proposed high-level healthcare data exchange model intended for Cross4all (C4A) project. After that, the used Use case scenarios for the C4A types of systems are defined. Concluding remarks summarize the aims of the project and explain the C4A goals.

2 Healthcare Data Exchange

In a centralized EHR patient data are collected and stored in a single repository or location. Individual healthcare providers or health professionals maintain the full details for individuals (patients) in their own EPR or EMR, which are subsets of the EHR.

In the distributed model of EHR, there is no location that is considered as a primary repository of information. Instead, the EHR is physically distributed across several locations. Users create their own view or record by accessing the data from these locations. In this model of EHR, each healthcare organization has its own EHR with its own data model and with its own terminology standards.

In some healthcare systems, EHR's data are distributed in the different IT systems in private and public healthcare institutions in a form of non-necessarily standardized EHR, EMR or EPR. The data exchange between IT systems designed in this way, is very difficult. Hence, the introduction of standards in the design of the new model of EHR is needed. There are different health informatics standards to define models of EHR: ISO HL7 21731 'Health informatics – HL7 version 3 Reference Information Model' (RIM) [16], ISO EN 13606 'Health Informatics – EHR communication' [10], Open - EHR Reference Model [17], etc. On the other hand, International Statistical Classification of Diseases 10th revision (ICD10) is used in most European EHRs.

Design of health information systems, especially their underlying business and informational models which describe basic concepts, business and relation networks, have to be based on standards. Lopez and Blobel [18] present a method and the necessary tool for reusing standard of healthcare information models, for supporting the development of semantically interoperable systems and components.

Yet, it seems that traditional EHRs, which are based on the 'fetch and show' model, provide limited functionality hardly covering the whole spectrum of the patients' related needs. Therefore, new solutions such as the personal-EHRs (PHRs) emerged to narrow this gap. In more detail, PHRs' data can come from various sources like EHRs, health providers (e.g., e-Prescription, e-Referral), and/or directly from the patient him/herself – including non-clinical information (e.g., exercise habits, food and diet statistics, probably gained with usage of HIS for screening of vital live signs, exposome data etc.) [19, 20].

The proposed model is differentiated from past efforts to integrate existing eHealth systems and sources of patient data in two key aspects. First, the information held and shared about a user is not restricted to medical or disease related data created in health structures to which the user is connected. It may well include also information beyond his/her medical history, such as information related to maintaining a healthy life, such as lifestyle and living conditions, altogether forming a more holistic representation of the person's health lifecycle, covering all kinds of physical, psychological, and social aspects. Secondly, in this PHR-based approach the records stored are intended to be fully owned and controlled by the user (patient) himself. In this way, it lies upon the user alone to decide who, when and for how long, will be granted access to what information and in which form or level of detail. The data and records of an individual in the project's model, being manually entered or automatically imported from a series of connected, authorized by the user, health providers and systems, are completely independent from third party entities and are not meant to replace, under any circumstances, the official records of other source providers.

The proposed model for the management of PHR data based on distributed cloud-based solution and storage and secure exchange is shown at Fig. 1.

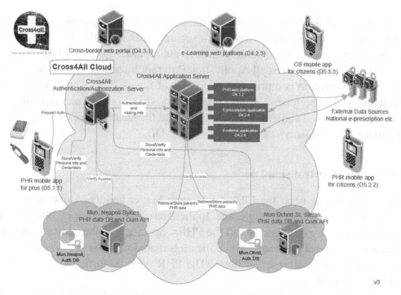

Fig. 1. Cross4all healthcare data exchange model

This project combines state-of-the-art techniques, including cloud computing and service-oriented architectures, thereby facilitating the integration of diverse electronic services and features that may support patient-centered care, telecare, self-care and overall better health data management and medical practice [21]. The project intends to include several applications and points-of-access for promoting better management and use of citizens' PHR, including cross-border use, with particular focus on serving the needs of elderly and people with disabilities [22] and socially/geographically isolated individuals. This integrated ecosystem of applications and features, as a whole, provides a wide range of capabilities, as it may:

- allow citizens to create and maintain their own ePHRs (optionally, with the support of the staff of the centers)
- allow citizens to import data to their own ePHR, automatically from connected/supported devices
- allow citizens to share data of their ePHRs, in a user-friendly manner, with health professionals across the borders
- allow automatic (i.e., easy and free of error) importing of data from third party systems that it interfaces with (e.g., existing healthcare electronic systems of various healthcare authorities, including national prescription systems, hospital electronic systems, private telecare services, etc.)
- offer info about & access to ePHR, eLearning, e-Prescription and e-Referral
- offer access to eLearning for citizens (health & digital health literacy)
- offer information on services, including information about access and accessibility of health infrastructures and services in their area
- support the operation of a volunteer programs for informal caregivers
- offer access to educational resources for health professionals

The backbone of the C4A ecosystem is the project's cloud, responsible for hosting the web based PHR application as well as for securely managing the data of each PHR. Security by design approach is exhibited through autonomous distributed databases and secure data connection between all Application Program Interfaces (APIs) and applications. The Cloud solution comprises of two types of nodes: The central node and the peripheral nodes. The peripheral nodes correspond to the pilot sites participating in the project and host the local authentication database and the health data DB along with the corresponding API web-services for managing that DB.

2.1 The Project Central cloud node

The central cloud node hosts the C4A Application Server and the Authentication and Authorization Server (Auth Server). The applications and repositories hosted in this cloud infrastructure are common for all the PHRs. Specifically, the central cloud node hosts two servers offering services and application shared to all project's users.

The project's application server hosts: The PHR web platform and e-prescription and e-referral applications.

The project Auth Server host *the central auth database* and *the authentication and authorization web API*. In this, the central auth database will be common for all project users' database where two types of data will be stored: a) The user credentials, i.e. unique id and password (valid email or a mobile telephone number) and b) Details on the location of actual health record data such as the server IP and identifier of the record in the corresponding peripheral cloud node.

The authentication and authorization web API will define a set of web services that will perform registration and authentication procedures and can be accessed either by the PHR web platform or by the non-web-based applications (i.e. PHR mobile app for pros and PHR mobile app for citizens).

2.2 The project peripheral cloud node

A peripheral cloud node instance will be deployed per pilot site. These nodes can be hot plugged i.e. additional peripheral nodes can be added without impacting existing operation, also existing peripheral nodes can be removed without rendering the rest of the system unusable. Of course, in the latter case data stored on the specific peripheral node will be inaccessible but the system will continue to operate normally and users with data on other peripheral nodes will be unaffected.

Each peripheral node hosts includes an instance of:

- Local authentication database (db) (In this db private data of the specific pilot study's participants are stored based on the requirements of the pilot site (like full name, date of birth, address or social security number). This db is managed only by the project Auth Server (i.e. PHR web platform or other applications do not have direct access));
- Health data DB and API (In this DB the health-related data of each pilot site's participant are stored and exposed through a secure web API. This API will allow authenticated applications to view/store data for the hosted participants. This means that PHR web platform and PHR mobile apps for pros and citizens upon authentication of the user they can store or retrieve data about him/her by directly accessing this API). In the project, two peripheral nodes will be deployed. But the architecture is seamlessly scalable by adding extra peripheral nodes with minimum configuration (i.e. register the new IP address to the central node).

2.3 Data model

The data model of the health data DB is based on HL7 FHIR resources. The communication with the backend model and data and metadata storage is done via a RESTful interface with resources representations' encoded in the Turtle 1.1 RDF format.

The authentication of the system is comprised of federation of Auth servers which run the open source Keycloak server. The Auth process is the following: when the external user wants to login to the PHR system it is redirected to login page in the central Keycloak server from which it is authenticated via username and password and if authentication is successful then the user is redirected to the appropriate national PHR web dashboard. The user data in the Auth servers is stored in instances of the Keycloak server in the respective country and based on the domain name of the username the user is first authenticated in one national server and if the user is not found on that server it is searched and authenticated in the other national server of the Keycloak federation. The Auth server does not store any medical or personal or identification data apart from data needed for authentication and reference to the patient id in the PHR itself.

Following proper authentication, a number of external systems can store or query data from or to the PHR. An integration with the National e-health system of Republic of North Macedonia (RNM) called "Moj termin" is designed in which a properly authenticated healthcare practitioner can export medication requests (with complete HL7 metadata) and import it in the PHR for their appropriate patient, also user of the PHR system. Similar integration is planned for exporting or "connecting" the referral information and doctor report document (as a result from the referral) to be imported and stored as HL7 entity in the PHR itself and to be in possession of the patient itself.

This DB model enables different stakeholders to perform different functions:

- Healthcare practitioners are able to discover a patient with at least one identification. By selecting the desired patient, they can submit the patient's EHR access request. Based on the authorization result and allowed access by the patient, the request is either allowed or denied;
- Patients are able to view their EHR from particular healthcare providers they are associated with or the composite EHR aggregated from all healthcare providers they obtained services from; and
- Administrators have the capability to manage all users, healthcare providers and other registered institutions, insurance companies, registered in the whole system.

3   Use Case Scenarios

The Cross4all project ecosystem consists of the following four integrated and interconnected components:

a) a Web portal and a connected mobile application for citizens, which:

   (1) offer information on available services, including information about access and accessibility for persons with disabilities
   (2) provides a facility for seeking support offered on a volunteer basis from local informal care givers
   (3) offer access to educational materials, guides and other resources for professionals
   (4) offer information about and access to the integrated ePHR, eLearning, e-Prescription and e-Referral services

b) an eLearning module for citizens, which is aimed to improve their health and digital health literacy
c) an PHR and a connected mobile app for citizens, which

   (1) allow citizens to create and maintain their own PHRs (optionally, with the support of external support experts)
   (2) allow citizens to import data to their own PHR, automatically from connected/supported devices
   (3) allow citizens to share data of their PHRs, in a user-friendly manner, with health professionals and providers, including across the borders

d) interfaces for connecting external third-party systems to the project's PHR, such as national e-prescription and e-referral systems, or private telecare systems or EHRs, so that data recorded in such systems can be automatically (i.e., easily and free of error) imported to the PHR owners of the project.

The concept of exposome [19] can be also taken into consideration because includes all traditional measurements in healthcare, known as bio-monitoring methods, as well as new types of measurements using sensors. The IoT concept includes a variety of wearable sensors connected to the computers or smart phones [4]. There is no limit for development sensors as measuring instruments that provide data such as spatio-temporal data with previous settings of measurement units, default values and the value ranges and some important environmental data that influence on human health (Fig. 2).

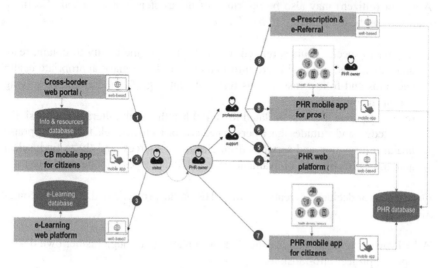

Fig. 2. Overview of the C4A subsystems and supported usages

In particular, the whole project ecosystem supports nine main use cases for eight main stakeholders as follows:

1. A Visitor (citizen or health professional or volunteer) can access through his own device or through a public project info-kiosk the Public web portal to:

   – find information on healthcare services, including information about access and accessibility, including maps, inspections data, transportation info, etc.
   – get involved into the volunteer program (view or announce or reply to request/offers for support services, etc.)
   – find educational resources produced in C4A project
   – find information on survey results and other key project outcomes
   – find/request information/guidance about existing tools, services and resources (project Help desk)

2. A Visitor (citizen or volunteer) may also download on his/her own mobile device and use the mobile app for citizens and access on the move information and geolocated guidance:

- search for information on healthcare services, including nearby options
- acquire personalized and georeferenced guidance and accessibility information (throughout their journeys across the region)
- make his own travel plan
- submit inquires to and receive replies from the Help Desk
- receive and send notifications related to the volunteer program
- receive mass notifications/alerts related to health tourists

3. A Visitor (citizen) may also be informed of the existence of a special eLearning platform and be redirected to it in order to:

- register to special courses related to health literacy (the ability to obtain, read, understand, and use healthcare information in order to make appropriate health decisions and follow instructions for treatment, e.g., for addressing or solving various health problems.
- register to special courses related to digital health 1 health literacy (set of skills, knowledge, and attitudes that a person needs in order to (a) seek, find, and appraise health information and services from electronic sources, and (b) to find, select and make effective use of available tools (PHRs, devices, mobile apps, etc.)

Once a Visitor decides to create his own PHR on the project cloud, he/she becomes a PHR owner.

4. A PHR owner (citizen) can use the PHR web platform through his/her own device or a public project info-kiosk to:

- store/view his/her own health profile and history
- store/view his/her own medication plan
- store/view his/her visits to health service providers and upload related medical documents (per visit)
- store/view his/her vital signals collected with his/her own (not connected) medical devices
- find, select and export data of his/her PHR in various formats
- find, select and share data of his/her PHR in various ways, e.g., with health professional within or across the boarders

To further support the management of citizen's PHRs, the concept of Local Support/Reference Centers is introduced, which shall help the citizens to:

- improve their self-awareness and self-management
- increase their awareness and about the project's tools/services
- provide user/technical support to system users
- generate and maintain an accurate PHR (health logistics)

These Centers shall be staffed with experts and shall be equipped appropriately, in order to be in the position to:

- provide information about available healthcare infrastructures and services at both sides of the border, including access and accessibility information
- act as an information center for citizens and help desk for medical tourists
- offer personal support for PHR management onsite (for active citizens)

offer personal support for PHR management at the citizen's space (for socially and geographically isolated individuals), though a pilot mobile units programme (Fig. 3).

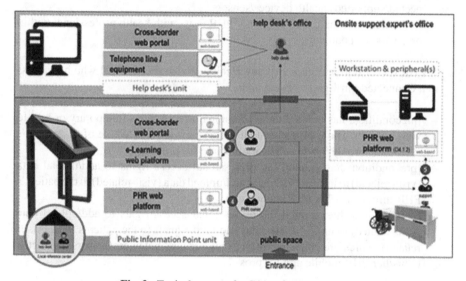

Fig. 3. Typical set-up of a C4A reference center

5. A PHR owner (citizen) may visit any Local support expert (in special municipal centers established in the project) to:

- receive assistance for learning how to use/manage his/her own PHR
- receive assistance for initiating/configuring his/her own PHR
- receive service for digitizing paper-based records of his/her and/or grant access for uploading them appropriately on his/her own PHR on his/her behalf
- Be informed of, and approve or not, data entries by the support staff

6. A PHR owner (citizen) may visit any Health service professional (who has no access to a connected third-party system) and:

- grant him/her access (temporary or not) to view parts or whole of his PHR, using the PHR web platform
- grant him/her access (visit-related) to add/upload data to his/her own PHR, using the PHR web platform
- Be informed of data viewed by the professional

    – Be informed of, and approve or not, data edited/entered by the professional

7. A PHR owner (citizen) download on his/her own mobile device the PHR mobile app for citizens or use a mobile set of devices provided by Cross4all to:

    – store new data to and view his/her own PHR
    – schedule and receive notifications related to his medical plan, doctor visits, etc.
    – store/view his/her vital signals collected with his/her own or provided (connected/supported) health devices/sensors
    – find, select and share data of his/her PHR, e.g., with health professional within or across the boarders

8a. A PHR owner (citizen) may visit any Health service professional who has access to a Connected third-party system and:

    – [precondition: has granted the professional access rights (temporary or not) to view parts or whole of his/her PHR data] view parts or whole of the patient's PHR data through his/her PHR mobile app for pros
    – [precondition: has granted the professional access rights to add/upload data (visit-related) to his/her PHR data] add/upload data (visit-related) to the patient's PHR through his/her PHR mobile app for pros
    – [precondition: has granted the professional access rights to add/upload data (visit-related) to his/her PHR data] add vital signals to the patients PHR (visit-related), which are collected through his/her health devices/sensors connected to his/her PHR mobile app for pros

8b. A PHR owner (citizen) may visit any Health service professional who has access to a Connected third-party system and (Fig. 4):

    – [precondition: has granted the professional access rights (temporary or not) to view parts or whole of his/her PHR data] view parts or whole of the patient's PHR data, through his/her e-Prescription & e-Referral systems
    – [precondition: has granted the professional access rights (temporary or not) to add/upload new data to his/her PHR data] add/upload data to the patient's PHR through his/her e-Prescription & e-Referral systems

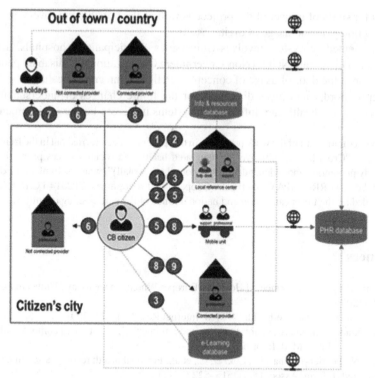

Fig. 4. Overview of access & use of systems and services, in and out of borders

4 Conclusions

Cloud Computing approach in combination with international healthcare standard in the EHR design enables healthcare data exchange among many healthcare organizations. In addition, it leads to the healthcare service quality improvement by strengthening the users' role in managing their own healthcare.

The C4A project aims to setup the bases for realization of the concept of self-ownership of PHR data, creating model with a powerful structure of Cloud based solution which will be created according to high level health care standards for patients' data security, no matter if they origins from EU countries or partner country. They will have a common distributed database with intelligent system of patient recognition as well as knowledge base empowered with information for a wide range of citizens and scenarios for increasing the e-health and digital health literacy for health services, no matter they are in the living country or abroad. This knowledge base will be accessible for the citizen as e-learning for e-health and health digital literacy system with a special focus for disable people for which the project provides accessible digital assets, applying Web Content Accessibility Guidelines (WCAG) standards for accessibility and availability.

The project implementation has to be done in cross border area and tested in the real environment in the intended pilot centers which have to provide a practice of proposed strategy, action plan and usage of digital assets cross borders. These pilot centers will the bases for study of impact of the project as well as the possible improvement on the strategy, action plan and usage of digital assets.

The presented ecosystem involves Universities, municipalities, hospitals, medical practitioner, patients and NGOs (Non-Government Organizations) of disabled people. It intends to open the door of usage of concept of PHR and integrated health and e-health system cross border and gives directions for the future changes of understanding of health services and health care information systems in a cross-border environment.

Acknowledgement. Part of the work presented in this paper has been carried out in the framework of the project "Cross-border initiative for integrated health and social services promoting safe ageing, early prevention and independent living for all (Cross4all)", which is implemented in the context of the INTERREG IPA Cross Border Cooperation Programme CCI 2014 TC 16 I5CB 009 and co-funded by the European Union and national funds of the participating countries.

References

1. Mountford, N., et al.: Connected Health in Europe: Where are we today? University College Dublin (2016)
2. Eysenbach., G.: What is e-health? J. Med. Internet Res. **3**(2), e20 (2001)
3. Lu, Y., Xiao, Y., Sears, A., et al.: A review and a framework of handheld computer adoption in healthcare. Int. J. Med. Inform. **74**(5), 409–422 (2005)
4. Archer, N., Fevrier-Thomas, U., Lokker, C., et al.: Personal health records: a scoping review. J. Am. Med. Inform. Assoc. **18**(4), 515–522 (2011)
5. Fisher, B., Bhavnani, V., Winfield, M.: How patients use access to their full health records: a qualitative study of patients in general practice. J. R. Soc. Med. **102**(12), 539–544 (2009)
6. Ralston, J.D., Martin, D.P., Anderson, M.L., et al.: Group health cooperative's transformation toward patient-centered access. Med. Care Res. Rev. **66**(6), 703–724 (2009)
7. Weppner, W.G., Ralston, J.D., Koepsell, T.D., et al.: Use of a shared medical record with secure messaging by older patients with diabetes. Diabetes Care **33**(11), 2314–2319 (2013)
8. Bergmo, T.S.: How to measure costs and benefits of eHealth interventions: an overview of methods and frameworks. J. Med. Internet Res. **17**(11), e254 (2015)
9. Bashshur, R.L., Shannon, G., Krupinski, E.A., et al.: Sustaining and realizing the promise of telemedicine. Telemed. J. e-Health **19**(5), 339–345 (2013)
10. ISO/DTR 20514:2005: Health Informatics-Electronic Health Record- Definition, Scope, and Context (2005). http://www.iso.org/iso/iso_catalogue/catalogue_tc/catalogue_detail.htm?csnumber=39525
11. Metzger, M.H., Durand, T., Lallich, S., et al.: The use of regional platforms for managing electronic health records for the production of regional public health indicators in France. BMC Med. Inform. Decis. Mak. **12**, 28 (2012)
12. Fernández-Cardeñosa, G., De la Torre-Díez, I., López-Coronado, M., et al.: Analysis of cloud-based solutions on EHRs systems in different scenarios. J. Med. Syst. **36**(6), 3777–3782 (2012)
13. Fernández-Cardeñosa, G., De la Torre-Díez, I., Rodrigues, J.J.P.C.: Analysis of the cloud computing paradigm on mobile health records systems. In: Proceedings of the Sixth International Conference on Innovative Mobile and Internet Services in Ubiquitous Computing, Palermo, Italy, July 2012, pp. 927–932. IEEE Computer Society Washington, DC, USA (2012)

14. Savoska, S., et al.: Design of cross border healthcare integrated system and its privacy and security issues. Comput. Commun. Eng. **13**(2), 58–63 (2019). First Workshop on Information Security 2019, 9th Balkan Conference in Informatics (2019)
15. Rodrigues, J.J.P.C., de la Torre, I., Fernández, G., et al.: Analysis of the security and privacy requirements of cloud-based electronic health records systems. J. Med. Internet Res. **15**(8), 186–195 (2013)
16. HL7 Inc. (2014) ISO HL7 21731 'Health informatics – HL7 version 3 Reference Information Model' (RIM) (2015). http://www.iso.org/iso/iso_catalogue/catalogue_tc/catalogue_det ail.htm?csnumber=61454. Accessed July 2015
17. Beale, T.: The open EHR Archetype Model (AOM) version 1.0.1 2007 (2015). http://www. openehr.org/releases/1.0.1/architecture/am/aom.pdf. Accessed June 2015
18. Lopez, D.M., Blobel, B.: Enhanced semantic interoperability by profiling health informatics standards. Methods Inf. Med. **48**(2), 170–177 (2009)
19. Savoska, S., Ristevski, B., Blazheska-Tabakovska, N., Jolevski, I.: Towards integration exposome data and personal health records in the age of IoT, ICT innovations 2019, Ohrid (2019)
20. Kulev, I., et al.: Development of a novel recommendation algorithm for collaborative health: care system model. Comput. Sci. Inf. Syst. **10**(3), 1455–1471 (2013)
21. Savoska, S., Jolevski, I.: Architectural model of e-health PHR to support the integrated cross-border services. In: Proceedings of ISGT 2018, Sofia (2018). http://ceur-ws.org/Vol-2464/ paper4.pdf
22. Blazheska-Tabakovska, N., Ristevski, B., Savoska, S., Bocevska, A.: Learning management systems as platforms for increasing the digital and health literacy. In: ICEBT 2019: Proceedings of the 2019 3rd International Conference on E-Education, E-Business and E-Technology, August 2019, pp. 33–37 (2019). https://doi.org/10.1145/3355166.3355176

Time Series Anomaly Detection with Variational Autoencoder Using Mahalanobis Distance

Laze Gjorgiev$^{(\boxtimes)}$ and Sonja Gievska

Faculty of Computer Science and Engineering, University of Sts. Cyril and Methodius in Skopje, Rugjer Boshkovikj 16, Skopje, Republic of North Macedonia
laze.gjorgiev@students.finki.ukim.mk, sonja.gievska@finki.ukim.mk

Abstract. Two themes have dominated the research on anomaly detection in time series data, one related to explorations of deep architectures for the task, and the other, equally important, the creation of large benchmark datasets. In line with the current trends, we have proposed several deep learning architectures based on Variational Autoencoders that have been evaluated for detecting cyber-attacks on water distribution system on the BATADAL challenge task and dataset. The second research aim of this study was to examine the impact of using Mahalanobis distance as a reconstruction error on the performance of the proposed models.

Keywords: Time series analysis · Anomaly detection · Variational autoencoder · Mahalanobis distance

1 Introduction

The ability to draw valid inferences from data has critical importance for a number of applications and industrial sectors, from mobile networks to medical diagnosis to manufacturing control systems. Malicious attacks (e.g., intrusions, fraudulent behavior, cyber-attacks), abnormality (e.g., malignant cells in images, irregular system behavior) and faulty measurements (e.g., sensors, industrial equipment, medical devices) however, pose a serious threat to the validity of the assertion drawn from data. Detecting and labeling anomaly events that are sparse, context-specific, difficult to define or distinguished from what is considered normal data are far from straightforward. Addressing these longstanding concerns have led to further advances in anomaly detection that has been regarded as an active area of research that complements predictive and prescriptive analytics.

Research work on the topic expanded from traditional approaches relying on classification, distance-based and statistical modeling [1–11] to exploiting the advances of diverse deep learning architectures [12,13]. New lines of research using deep generative neural networks [14,15] are being hailed for their potential for unsupervised learning from multivariate data with spatial and temporal

© Springer Nature Switzerland AG 2020
V. Dimitrova and I. Dimitrovski (Eds.): ICT Innovations 2020, CCIS 1316, pp. 42–55, 2020.
https://doi.org/10.1007/978-3-030-62098-1_4

dependencies. As with past imports in other areas, deep learning carries both great promise and substantial problems that are far from straightforward. From selecting the right deep architecture to learning its parameters, from scaling up the algorithms to understanding and interpreting model behaviors. These problems are faced by researchers and practitioners in the field that still awaits for robust and innovative guidelines and solutions.

Detecting cases of intrusion and anomaly in data observations entails 1) learning data representations of normal data i.e., mapping high-dimensional input data into a low-dimensional latent space, and 2) selecting the measure and threshold that will be used for distinguishing anomaly in the latent space. Handling high-dimensional data with spatial and temporal dependencies is pointed out as a major challenge for traditional approaches for anomaly detection, especially those inclined toward classification and clustering approaches. Principle Component Analysis (PCA) [11] is usually excluded, although the problem of hard-to-interpret anomaly data remains.

Extracting nonlinear and hierarchical discriminative features from complex high-dimensional data generated from IoT devices or real-time streaming data usually implies massive quantities of data that is sequential in nature and manifest characteristics such as trends and seasonality. Various deep learning architectures are suggested to handle the challenge. Convolutional Neural Networks (CNN) are considered better choice for capturing spatial and temporal dependencies when compared with the capabilities of Multilayer perceptron (MPA), while Recurrent neural networks (RNN) have been preferred for modeling sequence data. A special kind of RNNs, Long Short Term memory (LSTMs) are regarded particularly suited for capturing long-term dependencies in time-series data. Research on anomaly detection has established the central role deep learning plays in the success of a number of case studies [16–19]. Chalapathy and Chawla [13] present a comprehensive study of deep learning techniques for detecting anomalies and intrusions and highlights the performance advantages of deep learning models when faced with massive quantities of data.

Research interest in deep generative networks for anomaly detection, in particular Variational Autoencodeers (VAE) [14], has been intense in recent years. Variational Autoencoders are particularly useful for generating encodings of input data and then use them to re-construct its input, which in the case of anomaly detection corresponds to denoting a data sample that deviates from what is considered normal as the one that cannot be reconstructed. In addition, the anomaly scores and decision thresholds are probabilistic with a statistical meaning.

Following these current trends in deep learning for anomaly detection [16–19], we have evaluated the performance of various deep learning models using Variational Autoencoders on the Battle of the Attack Detection Algorithms – BATADAL[1] challenge task, which was introduced and extensively described in [20]. The second objective of this paper was to investigate the potential

---

[1] https://www.batadal.net/data.html.

performance gains if Mahalanobis distance measure [21] was used instead of mean square error for detecting data that deviates from normal.

In the next section, we provide a brief historical overview of traditional methods comparing them to the current trends in anomaly detection. We then give a description of the BATADAL dataset used in our experiments. In the fourth section, the Variational Autoencoders are presented, which is a basic component in the proposed deep models, followed by a discussion of the ten hierarchical deep architectures that were evaluated on the task at hand. We assess the current and potential contributions of the proposed models in the fifth section. The final sections concludes the paper and proposes a research agenda for future explorations.

2   Related Work

Significant number of methods have been explored over the last decades indicating the longstanding interest in the field of anomaly detection. In this short overview, we place emphasis on recent studies using deep learning for anomaly detection, but we also want to give reference to traditional approaches that still have a place in today's research.

Early work on anomaly detection tend to originate from machine learning standpoint toward detecting anomalies. Supervised or semi-supervised classification methods such as SVM [1], S3VM [2] or decision tree based models [3–5] such as Random Forrest [4,5] were used to distinguish between normal and outlier class. Furthermore, unsupervised methods such as proximity-based methods [6–8] (e.g. Nearest Neighbor [6], K-means [6], Lof [8]), angle-based methods [9], Isolation Forrest [10] or PCA [11] were used to separate normal from abnormal data. The premise in statistical modeling is that normal data occur in high-probability regions of the selected distribution (e.g., Gausian Mixture Model) that closely resembles the realistic data distribution. The major limitations of the traditional methods is their inability to handle massive quantities of complex high-dimensional data including those with dependencies.

Consistent with the accomplishments of deep learning for predictive analysis in a number of areas, from image processing to language understanding, various hierarchical deep neural networks have been used to learn nonlinear and discriminative indicators in multi-dimensional data that facilitates the task of anomaly detection. Two types of deep learning for anomaly detection are distinguished, prediction and reconstruction based. Prediction inclined ones are used to predict a point in future and detect anomaly based on the difference between the actual and predicted value i.e., prediction error. Different contexts and applications are likely to afford different neural architectures. DeepNap, Kim et al. (2018) [16] uses an AE with LSTM for prediction, and fully-connected MLP for detecting an anomaly in time-series data, using partial reconstruction error to train both modules jointly, while LSTM neural networks and multivariate Gaussian distribution as a prediction error have been considered a more suitable choice for anomaly detection in time series data [17].

Reconstruction-based techniques have embraced the new generation of deep generative networks, such as Variational Autoencoders (VAE) [14] and Generative Adversarial networks (GAN) [15] that have made possible the task of non-linear dimensionality reduction and learning representation of input data in unsupervised manner. We have given preference in this review by tracing the studies that contribute to the body of works on use of deep generative networks for anomaly detection in time series data.

A Variational Bi-LSTM Autoencoder proposed by Pereira and Silveira (2018) [22] uses a special attention mechanism called Variational SelfAttention Mechanism (VSAM) to improve the decoding process for anomaly detection in solar energy time series data. The VAE-reEncoder architecture, composed of two Bi-LSTM encoders and Bi-LSTM decoder, proposed for detecting anomalies in time-series data by Zhang and Chen (2019) in [19] has also been inspirational for our work. Two scores were taken into account: 1) a reconstruction error of the decoder and 2) a reconstruction error calculated by comparing the latent vectors generated by the second and the first encoder. An and Cho (2015) [23] propose new method for computing anomaly score with reconstruction probability using Variational Autoencoders that motivated our interest in investigating the potential of using the Mahalanobis distance in our comparative analysis.

Of particular interest to the present work are the findings of seven research studies that have participated in the original BATADAL competition against which we compare our results. The approaches taken by the participant teams range from data-driven to model-based. The best-performing model on the challenge task was reported by Housh and Ohar (2017) [24] that have built a simulation model of the hydraulic processes in the water distribution system using the popular EPANET [25] toolkit for modeling water distribution systems. Subsequently, the anomalies were identified by comparing the simulated values with the observed values provided by the dataset.

Abokifa et al. (2017) [26] introduced a three-stage detection method; the first one using statistical methods for detecting outliers, the second using a Multi-layered perceptron targeting contextual anomalies and the third one aimed at detecting global anomalies affecting multiple sensors by employing Principal Component Analysis as a dimensionality reduction method. Drawing on this experiences, we have also explored the idea of detecting contextual anomalies with neural networks, in our case it was a deep learning architecture instead of a Multi-layered Perceptron. The detection method reported in Giacomoni et al. (2017) [27], identifies the values below and above the thresholds associated with normal operation conditions for the water distribution system. Subsequently, Principal Component Analysis was used for the separation of normal data and those that are considered anomalies.

In an attempt to reduce the dimensionality and complexity of the problem, Brentan et al. (2017) [28], propose using an independent deep recursive neural network for each district (a division that was inherent to the C-Town water supply data) to predict the tank water levels as a function of other available features provided in the dataset. We have also explored this idea of independent detection across all districts and examined the impact on the performance of our models when compared to a more holistic approach of having one model for detecting the cyber-attacks. In another line of work, Chandy et al. (2017) [29] have adopted a two-phase detection model. First, any violations of the physical and/or operating rules are flagged as potential anomalies, which are latter confirmed by the second model, a Convolutional Variational Autoencoder (VAE) that calculates the reconstruction probability of the data. In our approach, we have also opted for the use of deep generative network, VAE, although we use hierarchical stacking of various layers (e.g., convolutional and LSTM) as opposed to a pure convolutional VAE structure that was used in their work.

Pasha et al. (2017) [30] presented an algorithm consisting of three main interconnected modules working on control rules and consistency checks, statistical pattern recognition for hydraulic and system relationships. A two-stage method extracts a four dimensional feature vector to capture the mean and covariance present in the multi-dimensional time series data, and latter uses Random Forest classifier to detect the cyber attacks was reported by Aghashahi et al. (2017) [31].

3   Dataset

The task of detecting cyber-physical attacks on water distribution system was the central objective posed by the competition challenge the BATtle of the Attack Detection ALgorithms (BATADAL) [20]. The accompanying dataset contained simulated observation data from the hydraulic monitoring system of a hypothetical C-Town medium-size water network. The participant were provided with two training and one test dataset, containing time series data sampled hourly from the simulated observations of 43 system variables divided into nine district metered areas.

The first training set contained system observations under normal operating conditions (no cyber attacks) for a period of 365 days. The second training dataset included 7 cyber attacks, one revealed and the other partially revealed or concealed from the participants. Detection algorithms were evaluated on a test dataset containing 7 additional concealed attacks, using the two scores described below.

Time-to-detection, denoted as S_{TTD}, measuring how fast an algorithm detects an attack was calculated as:

$$S_{TTD} = 1 - \frac{1}{N} \sum_{i=1} \frac{TTD_i}{\Delta_i} \tag{1}$$

where, N is the number of attacks, TDD_i is the time to detect (in hours) relative to the i-th attack and Δ_i is duration of an attack.

Classification performance score, S_{CLF}, assessing how accurate an algorithms is, was defined as:

$$S_{CLF} = \frac{TPR + TNR}{2} \tag{2}$$

where, TPR is true positive rate usually known as recall, and TNR is true negative rate usually known as specificity, which were calculated in a standard way based on the TP, TN, FP and FN.

These two evaluation scores were used in an overall ranking score S, defined as:

$$S = y * S_{TTD} + (1 - y) * S_{CLF} \tag{3}$$

The value of the coefficient was set to 0.5, giving equal weights to both evaluation metrics i.e. early and accurate detections.

We have performed some feature preparation by adding seasonal factors encoding derived from timestamps, such as sin and cos value of month-of-year, day-of-month, and hour-of-day; min-max normalization was subsequently performed for each seasonal dimension.

4 Methodology

4.1 Variational Autoencoders for Anomaly Detection

When faced with multivariate time series data, such as the BATADAL dataset, and being inclined toward exploring the potential of deep learning for anomaly detection, Variational Autoencoders [14] seemed an obvious choice. We give a brief overview of general VAE architecture shown in Fig. 1. Variational Autoencoder consists of two neural networks, encoder and decoder. The encoder maps high-dimensional input data x into a low-dimensional latent space z, while the role of the second neural network, decoder, is to generate or reconstruct the input data from the sampled latent vector z as reconstructed input data x'.

Calculating the posterior distribution $p(z \mid x)$ is not tractable, so variational inference is used to approximate p with some variational distribution $q(z \mid x)$; usually Gaussian distribution so that it can be feasibly calculated. Kullback-Leibler divergence (D_{KL}) is most commonly used to measure how close is one distribution to another, and in the case of VAE, it measures how close is the approximative posterior distribution $q(z \mid x)$ to the true prior $p(z)$. For calculation purposes, log-likelihood formulas are used, hence instead of minimizing $D_{KL}(q(z \mid x) \mid\mid p(z \mid x))$, the evidence lower bound ELBO, which is the lower bound of log-likelihood distribution of input data p(x) is maximized.

Variational Autoencoders try to maximize a stochastic objective function on training data given by the following equation:

$$ELBO(\theta, \psi; x) = E_{q_\theta}(z \mid x)log(p_\psi(x \mid z)) - D_{KL}(q_\theta(z \mid x) \mid\mid p(z)) \tag{4}$$

The first term, represents the reconstruction likelihood i.e. how good is the generative model at learning the training data, where $p_\psi(x \mid z)$ denotes the likelihood of obtaining the input vector x given the latent vector z, parametrized by the decoder neural network. The second term is the Kullback-Leibler divergence with a prior over the latent variables, where $q_\theta(z \mid x)$ is the approximated posterior distribution parameterized by the encoder neural network.

We estimate ELBO by sampling from q, so the "re-parametrization trick" in VAE is used. The encoder neural network that needs to learn the approximated posterior distribution $q(\mathbf{z} \mid \mathbf{x})$ takes the input $\boldsymbol{x}$ and outputs the parameters of $q(\mathbf{z} \mid \mathbf{x})$, e.g., a vector of means $\boldsymbol{\mu}$ and a vector of standard deviations $\boldsymbol{\sigma}$ for Gaussian distribution, from which $\boldsymbol{z}$ is sampled as an element-wise product of $\boldsymbol{\mu}$ and $\boldsymbol{\sigma}$ and passed to the decoder (see Fig. 1).

During training epochs, the first term of the objective function i.e., reconstruction likelihood, is calculated either in a traditional way as a mean square error or a Mahalanobis distance. Mahalanobis distance [21] takes into account the covariance between each pair of variables that exist in the space, therefore providing much more realistic and reliable way of measuring a distance between two points in a multivariate space. The equation for Mahalanobis distance between two points (X_a and X_b) that exists in n dimensional space with (nonsingular) covariance matrix Σ is the following:

$$Mh_{ab} = \sqrt{(X_b - X_a)^T * \Sigma^{-1} * (X_b - X_a)} \tag{5}$$

In our models, the Mahalanobis distance was computed between the original input vector $\boldsymbol{x}$ and the reconstructed input vector $\boldsymbol{x'}$ (averaged over 5 sampled vectors).

The anomaly threshold for every instance at time t is determined as follows:

$$|Mh_{xt}x'_t - \mu_n| > 3 * \sigma_n \tag{6}$$

where, $\boldsymbol{x}_t$ is the input vector at time t and $\boldsymbol{x'}_t$ is the averaged reconstructed input, μ_n and σ_n are the mean and standard deviation of the computed Mahalanobis distances for the previous n timestamps.

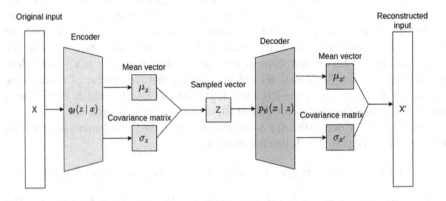

Fig. 1. Variational Autoencoder architecture

4.2   Models

Motivated by previous methodologies evaluated on the same dataset [26,28,29] that take into account the natural division of the C-Town water network into nine district metered areas, we have taken two modelling approaches. The first modelling approach takes a holistic view of the water distribution system operations. Modeling each of the nine district metered areas independently was used as an alternative approach, allowing us to reduce the dimensionality of the system time series data into nine components. In the second approach, the anomalies were detected per component and subsequently merged into a single set that was evaluated by the competition system. We have trained and tested several variants of Variational Autoencoders, the design decision were guided by previous research in the field. Each model has its complement, one modeling the entire water distribution system and the other modeling each time series originating from different district metered area – the area-based approach will be denoted by the letter C during acronym naming.

A short description of each of the ten architectures follows:

VAE-D. The encoder and decoder of this VAE variant are two neural networks, each represented by one dense layer. Each layer has 18 neurons, uses ReLU activation function and 5-dimensional latent space.

VAE-D-C. The simple architecture of VAE-D with two neural networks with dense layers was created for each time series component (corresponding to one district metered area). The number of units per layer and the dimensionality of the latent space were tailored to the specifics of the district metric area being modeled.

LSTM-VAE. This VAE variant has two LSTM networks as encoder and decoder in an attempt to capture the long-term dependencies when modeling sequences. The encoder is consisted of 1 LSTM layer with 18 units with ReLU connected to Dense layer with number of units to parametrize the 6 dim latent space. The input to the model is a window of 24 consecutive hours of data.

LSTM-VAE-C. This is a variant that uses one LSTM-VAE for each district metered area i.e., each of the 9 time series components.

LSTM-VAE-2E. Variational AutoEncoder with two encoders and a decoder. The first encoder has the simple architecture of VAE-D that generates the sampled encodings in the latent space. The second encoder is an LSTM neural network with two stacked layers that learns to predict an instance for the next timestamp given a sequence of previous 24 instances; the second LSTM layer outputs the final hidden state for the sequence. This final state is taken as a compression where all the relevant information are kept for the sequence in order to predict the next instance in the time series. Each LSTM layer has 10 neurons with activation ReLU and uses dropout. The input of these two encoders is concatenated and fed into a decoder. There is single model for the time series.

LSTM-VAE-2E-C. Same architecture as previously described LSTM-VAE-2E is used to model each of the 9 components, by adjusting the hyperparameters to the modeled system parameters.

CNN-VAE. Guided by the modeling approach taken in [29], we have explored the performance of a VAE variant that has an encoder, which is a Convolutional neural network (CNN) suitable to extract the local contextual relationships between the instances in our time series data. We use CNN with one convolution layer, Max pooling with filter size of 27 and kernel size of 6. The decoder has the same architecture as VAE-D.

CNN-VAE-C. This is component-wise variant of the previous model CNN-VAE, i.e., 9 CNN-VAE networks for each area with their hyperparameters selected to fit the problem being modeled.

VAE-ReEncoder. In this model, similarly as in [19], we use the encoder twice. In the first pass the encoder generates latent distribution from the input data that is sampled and passed through the decoder, so that in the second pass, the output generated by the decoder is sampled and fed to the input of the encoder. Two scores are taken into account for computing the anomaly score: 1) the Mahalanobis distance between the sampled reconstructed vector x' and the input vector x and 2) the mean square error between the latent vectors generated by the first and second pass through the encoder.

VAE-ReEncoder-C. Same as VAE-ReEncoder but here we have a different model for each component.

5 Discussion of Results

An exploratory study was conducted to assess the performance standing of ten hierarchical Variational Autoencoders architectures trained and evaluated on BATADAL datasets. The same evaluation criteria established for the BATADAL competition were used to compare the performance of the models, namely, the best overall score (S), time-to-detection (S_{TTD}) and classification performance score (S_{CLF}). Table 1 presents the evaluation metrics for each VAE model using the proposed Mahalanobis metric as a reconstruction loss. Surprisingly, the first model with the most simple architecture, VAE-D has obtained the best results on two metrics, overall score and the classification performance score. It seems that the model was able to capture the relationships between the 9 system components (areas), since its complemented architecture VAE-D-C that models each component separately underperforms on the same task.

The current findings demonstrate lower performance achieved by the LSTM variants of Variational Autoencoders, LSTM-VAE and LSTM-VAE-C, when compared with the best performing model. However, both of the models has shown almost perfect time-to-tetection score with LSTM -VAE-C being the model with the best TPR i.e. recall score. But the LSTM-VAE has failed to detect enough anomalies and yielded poor recall performances and LSTM-VAE-C has shown poor TNR i.e. specificity performances. We could hypothesized

Table 1. Evaluation metrics of the proposed deep learning models using Mahalanobis distance as a distance metric

Model	S	S_{TTD}	S_{CLF}	Precision	#Attacks detected
VAE-D	**0.800**	0.975	**0.624**	0.412	7
LSTM-VAE	0.735	0.979	0.491	0.185	7
LSTM-VAE-2E	0.664	0.820	0.508	0.236	7
CNN-VAE	0.523	0.543	0.503	0.227	5
VAE-ReEncoder	0.752	0.935	0.570	0.276	7
VAE-D-C	0.778	0.987	0.570	0.257	7
LSTM -VAE-C	0.778	0.999	0.556	0.216	7
LSTM-VAE-2E-C	0.761	**1.0**	0.523	**0.501**	7
CNN-VAE-C	0.713	0.931	0.496	0.191	7
VAE-ReEncoder-C	0.726	0.940	0.512	0.211	7

that the reason for the results lies in the complexity of the networks i.e., the dataset was not large enough to learn the high number of parameters needed to be learned.

The LSTM-based VAEs with 2 encoders have also shown lower performances, while the LSTM-VAE-2E-C beeing the model with best precision. The result could be contributed to the use of two encoders, the first one tasked with extracting high-dimensional features and relationship from the normal input observations, while the other encoder, was capable of extracting the temporal dependencies in a particular window of data using LSTM networks. The lack of large enough training set still accounts for lower performance compared with the simplest and best performing model VAE-D.

Little or no evidence has supported the advantages of using Convolutional variants of VAEs for cyberattack detection on the water distribution system, although CNN-VAE has yielded best TNR, but at the same time very poor score for TPR. One possible speculation for the results, could be attributed to the inadequacy between the size of the training dataset and the number of parameters needed to be learned. In order to capture all local and temporal dependencies from high dimensional time series the convolution layers needed better tuning. Meanwhile, the Conv-VAE-C version that independently models each district area, has shown results closer to the others, more likely because the convolutional networks were with lesser complexity i.e., the parameters were trained on partitons of the time series data.

The VAE-ReEncoder variants have similarly exhibited unsatisfactory performances. The reason could be the one previously discussed, the performance gains of deep learning networks are notoriously dependent on the amount of training data. In addition, more experiments for hyperparameters tuning are required to further advance the architectures.

One of the objective of this study, was to evaluate the impact of using Mahalanobis distance as opposed to the traditional distance metric, such as mean square error. We have conducted additional experiments for comparative analysis of the models and the results are displayed in Table 2. As hypothesized, the evaluation of the best performing models, VAE-D and VAE-D-C, revealed the significance of using a metric that is most suited to the task. It can be seen that both models have shown significant performance advantage when Mahalanobis distance is used, pointing to the suitability and relevance of using more appropriate measure to represent the loss in the distribution output of VAEs as well as a metric that takes into account the variance that exists in multivariate data.

Table 2. Comparison of the evaluation metrics of the best performing models using mean square error and Mahalanobis distance as a distance metric

Model	S	S_{TTD}	S_{CLF}	Precision	#Attacks detected
VAE-D	**0.800**	0.975	**0.624**	0.412	7
VAE-D-MSE	0.702	0.821	0.584	**0.533**	7
VAE-D-C	0.778	0.987	0.570	0.257	7
VAE-D-C-MSE	0.750	**1.0**	0.500	0.194	7

Table 3. Comparison of the evaluation metrics of our best performing model VAE-D with the models submitted to the BATADAL competition task

Rank	Team	#Attacks detected	S	S_{TTD}	S_{CLF}
1	Housh and Ohar	7	**0.970**	**0.975**	**0.953**
2	Abokifa et al.	7	0.949	0.958	0.940
3	Giacomoni et al.	7	0.927	0.936	0.917
4	Brentan et al.	6	0.894	0.857	0.931
5	Chandy et al.	7	0.802	0.835	0.768
6	VAE-D	7	0.800	**0.975**	0.624
7	Pasha et al.	7	0.773	0.885	0.660
8	Aghashahi et al.	3	0.534	0.429	0.640

Table 3 presents the performance results of our best model and the results reported by the 7 participant teams at the BATADAL competition. The superiority of our model in terms of time to detection of an attack is evident, although the VAE-D models fails to prevent many false positives and to detect all the instances that belong to an attack. It is interesting that it almost perfectly detects if there is an attack or in other words with almost perfect latency it detects the first instance of a series of instances during an attack, but it fails to detect the following instances in the ongoing attack. We would like to underline that no validation dataset was used for detecting anomalies, which might be a possible explanation for the lower performance compared to the best performing BATADAL models. We differ the discussion on the impact the cut-off thresholds for defining outliers has on the performance, until a detailed analysis is done.

6 Conclusion and Future Work

The present research explores several variants of Variational Autoencoders on the task of detection of attacks on a simulated observations in a water distribution system. The premise of the proposed methods for unsupervised deep learning anomaly detection is that attacks i.e., anomalies are events that cannot fit perfectly with the model operating under normal conditions. Variational Autoencoders use two neural networks, encoder and decoder, to model and learn the normal data. Our best performing model showed almost perfect results in time to detection of intrusions that are sparse and never encountered before without using any domain-specific knowledge during training. The effect of using the Mahalanobis distance as a reconstruction error is promising although needs further investigation. Future research directions will focus on increasing the model performance standing towards batter recall and sensitivity. We continue to experiment with hyperparameter tuning and transfer learning on various benchmark datasets. Investigations of advanced techniques, such as attention mechanism and self-supervised learning will also be explored. Taken together, our exploration study fits into the line of research on deep generative learning for anomaly detection.

Acknowledgement. This work was partially financed by the Faculty of Computer Science and Engineering at the "Ss. Cyril and Methodius" University.

References

1. Chandola V., Banerjee A., Kumar V.: Anomaly detection: a survey. ACM Comput. Surv. July 2009. Article No. 15. https://doi.org/10.1145/1541880.1541882
2. Chapelle, O., Chi, M., Zien, A.: A continuation method for semi-supervised SVMs. In: Proceedings of the 23rd international conference on Machine learning (ICML '06), pp. 185–192. Association for Computing Machinery, New York (2006). https://doi.org/10.1145/1143844.1143868
3. Reif, M., Goldstein, M., Stahl, A., Breuel, T.: Anomaly detection by combining decision trees and parametric densities. In: 2008 19th International Conference on Pattern Recognition, Tampa, FL, pp. 1–4 (2008). https://doi.org/10.1109/ICPR.2008.4761796
4. Breiman, L.: Random forests. Mach. Learn. **45**, 5–32 (2001). https://doi.org/10.1023/A:1010933404324
5. Primartha R., Tama B. A.: Anomaly detection using random forest: a performance revisited. In: 2017 International Conference on Data and Software Engineering (ICoDSE) (2017). https://doi.org/10.1109/ICODSE.2017.8285847

6. Goldstein, M., Uchida, S.: A comparative evaluation of unsupervised anomaly detection algorithms for multivariate data. PLoS One **11**(4), e0152173 (2016)
7. Knorr, E.M., Ng, R.T.: Algorithms for mining distance-based outliers in large datasets. In Proceedings of VLDB 1998, pp. 392–403 (1998). https://doi.org/10.5555/645924.671334
8. Breunig, M.M., Kriegel, H.-P., Ng, R.T., Sander, J.: LOF: identifying density-based local outliers. In: Proceedings of SIGMOD 2000, pp. 93–104 (2000). https://doi.org/10.1145/342009.335388
9. Pham, N., Pagh, R.: A near-linear time approximation algorithm for angle-based outlier detection in high-dimensional data. In: Proceedings of the 18th ACM SIGKDD International Conference on Knowledge Discovery and Data Mining, KDD 2012, August 2012, pp. 877–885 (2012). https://doi.org/10.1145/2339530.2339669
10. Liu, F.T., Ting, K.M., Zhou, Z.-H.: Isolation forest. In: IEEE International Conference on Data Mining, 2009, pp. 413–422. https://doi.org/10.1109/ICDM.2008.17
11. Jolliffe, I.T., Cadima, J.: Principal component analysis: a review and recent developments. Philos. Trans. A Math. Phys. Eng. Sci. **374**(2065), 20150202 (2016). https://doi.org/10.1098/rsta.2015.0202
12. Munir, M., Dengel, A., Chattha, M.A., Ahmed, S.: A comparative analysis of traditional and deep learning-based anomaly detection methods for streaming data. In: 18th IEEE International Conference on Machine Learning and Applications (ICMLA) At: Boca Raton, Florida, USA, December 2019 (2019). https://doi.org/10.1109/ICMLA.2019.00105
13. Chalapathy, R., Chawla, S.: Deep learning for anomaly detection: a survey. arXiv preprint arXiv:1901.03407 (2019)
14. Diederik, P.K., Welling, M.: Auto-Encoding Variational Bayes. arXiv:1312.6114 (2013)
15. Goodfellow, I.J., et al.: Generative Adversarial Networks. arXiv:1406.2661 (2014)
16. Kim, C., Lee, J., Kim, R., Park, Y., Kang, J.: DeepNAP: deep neural anomaly pre-detection in a semiconductor fab. Inf. Sci. **457–458**, 1–11 (2018). https://doi.org/10.1016/j.ins.2018.05.020
17. Malhotra, P., Vig, L., Shroff, G., Agarwal, P.: Long short term memory networks for anomaly detection in time series. In: European Symposium on Artificial Neural Networks, Computational Intelligence and Machine Learning, 2015, pp. 89–94
18. Xu, H., et al.: Unsupervised anomaly detection via variational autoencoder for seasonal KPIs in web applications. arXiv:1802.03903 (2018)
19. Zhang, C., Li, S., Zhang, H., Chen, Y.: VELC: a new variational autoencoder based model for time series anomaly detection (2019). arXiv:1907.01702
20. Taormina, R., et al.: The battle of the attack detection algorithms: disclosing cyber attacks on water distribution networks. J. Water Resour. Plann. Manage. **144**(8) (2018) https://doi.org/10.1061/(ASCE)WR.1943-5452.0000969
21. Mclachlan, G.J.: Mahalanobis distance. Resonance **4**(6), 20–26 (1999). https://doi.org/10.1007/BF02834632
22. Pereira, J., Silveira, M.: Unsupervised anomaly detection in energy time series data using variational recurrent autoencoders with attention. In: 17th IEEE International Conference on Machine Learning and Applications (ICMLA), Orlando, Florida, USA, December 2018 (2018). https://doi.org/10.1109/ICMLA.2018.00207
23. An, J., Cho, S.: Variational autoencoder based anomaly detection using reconstruction probability. Spec. Lect. IE **2**(1), 1–18 (2015)

24. Housh, M., Ohar, Z.: Model based approach for cyber-physical attacks detection 547 in water distribution systems. In: World Environmental and Water Resources Congress 2017, vol. 548, pp. 727–736 (2017). https://doi.org/10.1061/9780784480625.067

25. Maruf, M., et al.: Water distribution system modeling by using EPA-NET software. In: International Conference on Recent Innovation in Civil Engineering for Sustainable Development (IICSD-2015), DUET, Gazipur, Bangladesh, December 2015 (2015)

26. Abokifa, A.A., Haddad, K., Lo, C.S., Biswas, P.: Detection of cyber physical attacks on water distribution systems via principal component analysis and artificial neural networks. In: World Environmental and Water Resources Congress 2017, vol. 495, 676–691 (2017). https://doi.org/10.1061/9780784480625.063

27. Giacomoni, M., Gatsis, N., Taha, A.: Identification of cyber attacks on water distribution systems by unveiling low-dimensionality in the sensory data. In: World Environmental and Water Resources Congress 2017, vol. 531, pp. 660–675 (2017). https://doi.org/10.1061/9780784480625.062

28. Brentan, B. M., et al.: On-line cyber attack detection in water networks through state forecasting and control by pattern recognition. In: World Environmental and Water Resources Congress 2017, vol. 510, pp. 583–592 (2017). https://doi.org/10.1061/9780784480625.054

29. Chandy, S.E., Rasekh, A., Barker, Z.A., Campbell, B., Shafiee, M.E.: Detection of cyber-attacks to water systems through machine-learning-based anomaly detection in SCADA data. In: World Environmental and Water Resources Congress 2017, vol. 520, 611–616 (2017). https://doi.org/10.1061/9780784480625.057

30. Pasha, M.F.K., Kc, B., Somasundaram, S.L.: An approach to detect the cyber-physical attack on water distribution system. In: World Environmental and Water Resources Congress 2017, vol. 584, pp. 703–711 (2017).https://doi.org/10.1061/9780784480625.065

31. Aghashahi, M., Sundararajan, R., Pourahmadi, M., Banks, M.K.: Water distribution systems analysis symposium; battle of the attack detection algorithms 501 (BATADAL). In: World Environmental and Water Resources Congress 2017, vol. 502, pp. 101–108 (2017). https://doi.org/10.1061/9780784480595.010

Towards Cleaner Environments by Automated Garbage Detection in Images

Aleksandar Despotovski[1], Filip Despotovski[1], Jane Lameski[2],
Eftim Zdravevski[1], Andrea Kulakov[1], and Petre Lameski[1(✉)]

[1] Faculty of Computer Science and Engineering, University of Ss. Cyril and
Methodius in Skopje, Ruger Boskovic 16, 1000 Skopje, North Macedonia
[2] Faculty of Informatics and Mathematics, Technical University of Munich,
Munich, Germany
lameski@finki.ukim.mk
https://www.finki.ukim.mk
https://www.tum.de/

Abstract. The environment protection is becoming, now more than ever, a serious consideration of all government, non-government, and industrial organizations. The problem of littering and garbage is severe, particularly in developing countries. The problem of littering is that it has a compounding effect, and unless the litter is reported and cleaned right away, it tends to compound and become an even more significant problem. To raise awareness of this problem and to allow a future automated solution, we propose developing a garbage detecting system for detection and segmentation of garbage in images. For this reason, we use deep semantic segmentation approach to train a garbage segmentation model. Due to the small dataset for the task, we use transfer learning of pre-trained model that is adjusted to this specific problem. For this particular experiment, we also develop our own dataset to build segmentation models. In general, the deep semantic segmentation approaches combined with transfer learning, give promising results. They show great potential towards developing a garbage detection application that can be used by the public services and by concerned citizens to report garbage pollution problems in their communities.

Keywords: Image segmentation · Environment protection · Deep learning · Deep semantic segmentation

1 Introduction

Littering and garbage pollution is one of the oldest problems in any human settlement. The increased garbage production is connected to the increased population and the increased development of the living, commercial, and industrial areas [4]. Garbage pollution and collection problems are especially emphasized in developing countries where the garbage collection services are overwhelmed.

V. Dimitrova and I. Dimitrovski (Eds.): ICT Innovations 2020, CCIS 1316, pp. 56–63, 2020.
https://doi.org/10.1007/978-3-030-62098-1_5

Furthermore, the lack of environmental awareness further increases garbage pollution, especially in areas with lower development. The process of garbage collection automation is already present in different developed areas in the world, and there are quite a few examples of this automation. In [13] authors use computer vision techniques to estimate the volume of the dump based on eight different perspectives with about 85% accuracy. Authors in [16] use a web camera to estimate the garbage quantity in garbage cans for a smart garbage collection system. In [11] authors use Arduino and ultrasound sensor to estimate how full are the garbage collection units and alert the driver which bins they should visit and empty. [14,15] use deep learning architecture for segmentation and classification of aerial scenes.

Different approaches are used to increase garbage collection and recycling efficiency. In [3], authors propose a gamification approach to motivate society to be more involved in garbage recycling. Their case study proved that gamification approaches increase recycling and involvement significantly. To increase the awareness and help the garbage collection services, the new developments in computer vision can be of significant aid. In [13] is presented the architecture behind a smartphone app that detects and coarsely segments garbage regions in geo-tagged user images. It utilizes deep architecture of fully convolutional networks for detecting garbage in images with sliding windows, achieving a mean accuracy of 87.69 on the GINI dataset.

In this paper, we use machine learning techniques to design a system that would detect and segment garbage in images. We try to resolve the problem of detection by using image segmentation and deep semantic segmentation methods to achieve as good segmentation for garbage in images as possible. A successful segmentation would allow us to not only detect the garbage in the images but also allow the system to find the exact location of the garbage and possibly determine the type of garbage.

The motivation behind this work is to build a system that would allow the users to automatically detect garbage in images and send the images to local and state authorities. The goal is to increase public awareness about misplaced garbage and garbage pollution in urban areas, especially in the developing countries where the legal authorities involvement and public awareness is on a superficial level. Such a system would ease the automation in garbage detection and increase the possibilities for applications that would allow reducing the littering problem in urban and suburban areas. Our approach can also be used to classify the types of garbage and with further development, estimate the quantity of garbage in the area based on a single image.

The main contribution of this paper is the novel dataset for garbage detection that contains labelled pixels of several garbage classes and the application of deep learning segmentation architectures to obtain initial models and results.

We organized the paper as follows: In Sect. 2, we describe the used dataset, the data augmentation methods and model architecture for our experiments. In Sect. 3 we present the obtained results and finally in Sect. 4 we shortly describe the conclusion of our work and the future work.

2   Methods

2.1   Dataset

One of the essential resources for building a machine learning model is the data. For garbage detection and segmentation in images, at the moment of performing the experiments, there were no publicly available dataset, to the best of our knowledge. Although there are many garbage images on the Internet, there were no adequate annotations for those images which mark the garbage segments at a pixel level. For this reason, we gathered a collection of 100 images from the Internet and our environment that contain garbage. The images are diverse; some of them capture pollution in nature, while others contain garbage on the streets. The images are in different sizes and resolutions. Before the labelling and training process, we resized all of the images to 640 × 480 pixels.

We manually segmented and labelled every image in the dataset into the garbage and non-garbage regions. The segmentation is needed so that the machine learning models can learn which parts of the images represent garbage and even to which garbage type. Some such examples where the garbage content is manually segmented are given in Fig. 3 and Fig. 1.

Additionally, we manually classified the segmented images to the seven classes: Non-garbage (NG), Plastic (PL), Metal (ME), Paper (PA), Glass (GL), Other (OT). Figure 1 shows one example of a labelled image from the dataset.

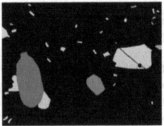

Fig. 1. An image from the dataset containing multiple class labeled segments and the corresponding labels.

The significantly lower number of pixels representing the garbage class compared to the non-garbage class presents another issue that increases the complexity of the problem. Namely, the dataset would be highly imbalanced, which is expected considering the nature of the problem. The percentages of each class are shown in Fig. 2.

To further increase the number of samples in the training set, we performed data augmentation. We used the Image Augmentation module implemented in the Keras [7] and Tensorflow [1] libraries. This approach transforms the images by slight rotation, scaling, zooming, adding noise, skewing, random rotation, horizontal and vertical translation, flipping and zooming to obtain new images

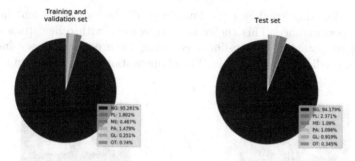

Fig. 2. Class distribution of the train (left) and test (right)

Fig. 3. An image from the dataset, next to its segmentation. The white segments represent garbage regions.

that are somewhat different than the original. We used data augmentation technique only for the training set. The remaining images represented the test set and were used to evaluate the trained models.

2.2   Models

For tackling the problem of detecting garbage in images, we used a deep learning approach, which utilizes deep neural networks for predicting the most plausible image segmentation.

Several recent state-of-the-art approaches commonly used for semantic segmentation, could be applied to this problem [2,5,10,12,18,19]. The success shown in several competitions, including [5], motivated us to choose the TernausNet deep learning architecture [10]. TernausNet is based on pre-trained VGG16 as an encoder linked with U-net [17] decoder layers. Similarly to the TernausNet, we combined VGG16 with deep CNN architecture SegNet [2]. The described model here is similar to [12], where it was applied for segmentation of skin lesions, but without using the pooling indices. For this experiment, we used a convolutional neural network with 13 convolutional layers combined in 5 blocks, each having a max-pooling and down-sampling layer. Additionally, we added 6 blocks of 2 convolutional layers with a dropout rate of 0.2 to regulate overfitting [19]. After each of the five blocks used to up-sample the output, another layer is added for concatenation of the transposed convolutional layer and the convolutional layer with the corresponding dimension, resulting in a so-called skip connection that combines the knowledge of the two levels and improves the model performance. This choice is in agreement with other studies where the combined potential of non-linear encoding features have shown increased accuracy in predictive tasks [6,8,9]. The segmentation model is shown in Fig. 4.

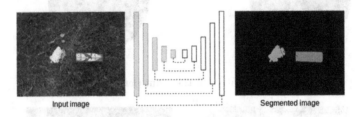

Input image   Segmented image

Fig. 4. Diagram of the fully convolutional neural network. The green blocks of layers are pre-trained and we only train the white layers. The dotted lines depict the skip connections between the layers (Color figure online)

The results in the next section were achieved using 300 epochs with the Adam optimizer and 0.01 learning rate for training the model. We used categorical cross-entropy loss function, which is usually used for segmentation and classification tasks.

3   Results

Based on the experimental setup described in the previous section, we trained the models and used the test set to evaluate them. For the garbage vs non-garbage segmentation, we obtained a DICE coefficient score of 0.62. As expected, the segmentation based on pre-trained deep learning yielded acceptable results. Figure 5 shows one example of the results to illustrate the effectiveness of the approach. The picture shows that the garbage is localized correctly and the segmentation successfully covers most of the garbage pixels. The DICE coefficient of this segmentation is 0.86, which is better than the result obtained for the whole dataset.

Fig. 5. Example result of the segmentation. Input image (left), Ground truth (middle), Obtained result (right)

Table 1. Evaluation of the segmentation for each label class

Segment class	Dice coefficient
Non-garbage (NG)	0.99
Plastic (PL)	0.58
Metal (ME)	0.34
Paper (PA)	0.58
Glass (GL)	0.25
Other (OT)	0.35

For the multi-label segmentation the results are shown in Table 1. The model is quite successful in segmenting the garbage from plastic and paper material. However, the most difficult to segment is the glass type of garbage is the most difficult to segment. The reason for this could be the lack of samples, considering that the glass material garbage class is under-represented in the dataset comparing to the top-scoring classes. A sample of the segmentation results is shown in Fig. 6.

Fig. 6. Example of segmentation results for glass-based garbage class on the test set. Original image (left), Ground truth segmentation (Middle), Segmentation result (right)

4   Conclusion

In this paper, we present a novel dataset for multi-label segmentation of garbage, and we tackle the problem of garbage segmentation. The primary motivation is to create a system that would allow automatic detection of garbage in images that can be used from the public organizations and citizens. For this purpose, considering the lack of available datasets, we have created our own dataset and used it to test several approaches. To increase the number of images, we used data augmentation, and we applied transfer learning to improve the model performance. The results show that our approach is suitable for garbage detection in images. However, it needs further improvements for detecting different types of garbage. Several ideas could be attempted in future works. One is to improve the dataset class balance because the current class imbalance causes an obstacle when training any machine learning model. Another idea is to add weights to the trainers and see if the models can be further improved.

Something that is missing in our experiments due to performance constraints, but could significantly improve the results is hyperparameter tuning. The learning rate, the number of training epochs, as well as the dropout rates were chosen on intuition and empirical results during the experiment. With proper hyperparameter optimization, it is reasonable to expect better results.

An additional idea is to increase the dataset size by non-standard augmentation techniques, such as, to use artificially created data using the placement of known segments over different, natural and artificial backgrounds, to increase the number of samples further. Obtaining and labelling additional images is always an option and could improve the performance.

Acknowledgments. The work presented in this paper was partially funded by the Ss. Cyril and Methodius University in Skopje, Faculty of Computer Science and Engineering. We also gratefully acknowledge the support of NVIDIA Corporation through a grant providing GPU resources for this work. We also acknowledge the support of the Microsoft AI for Earth for providing processing resources.

References

1. Abadi, M., et al.: TensorFlow: Large-scale machine learning on heterogeneous systems (2015). http://tensorflow.org/. Software available from tensorflow.org
2. Badrinarayanan, V., Kendall, A., Cipolla, R.: SegNet: a deep convolutional encoder-decoder architecture for image segmentation. IEEE Trans. Pattern Anal. Mach. Intell. **39**(12), 2481–2495 (2017)
3. Briones, A.G., et al.: Use of gamification techniques to encourage garbage recycling. A smart city approach. In: Uden, L., Hadzima, B., Ting, I.-H. (eds.) KMO 2018. CCIS, vol. 877, pp. 674–685. Springer, Cham (2018). https://doi.org/10.1007/978-3-319-95204-8_56
4. Brown, D.P.: Garbage: how population, landmass, and development interact with culture in the production of waste. Resour. Conserv. Recycl. **98**, 41–54 (2015). http://www.sciencedirect.com/science/article/pii/S0921344915000440
5. Carvana: Carvana image masking challenge automatically identify the boundaries of the car in an image. https://www.kaggle.com/c/carvana-image-masking-challenge/. Accessed 30 May 2019
6. Chen, L.-C., Zhu, Y., Papandreou, G., Schroff, F., Adam, H.: Encoder-decoder with atrous separable convolution for semantic image segmentation. In: Ferrari, V., Hebert, M., Sminchisescu, C., Weiss, Y. (eds.) ECCV 2018. LNCS, vol. 11211, pp. 833–851. Springer, Cham (2018). https://doi.org/10.1007/978-3-030-01234-2_49
7. Chollet, F., et al.: Keras (2015). https://github.com/fchollet/keras. Accessed 30 May 2019
8. Corizzo, R., Ceci, M., Japkowicz, N.: Anomaly detection and repair for accurate predictions in geo-distributed big data. Big Data Res. **16**, 18–35 (2019)
9. Corizzo, R., Ceci, M., Zdravevski, E., Japkowicz, N.: Scalable auto-encoders for gravitational waves detection from time series data. Expert. Syst. Appl. **151**, 113378 (2020)
10. Iglovikov, V., Shvets, A.: TernausNet: U-Net with VGG11 encoder pre-trained on ImageNet for image segmentation. ArXiv e-prints (2018)

11. Kumar, N.S., Vuayalakshmi, B., Prarthana, R.J., Shankar, A.: IoT based smart garbage alert system using Arduino UNO. In: 2016 IEEE Region 10 Conference (TENCON), pp. 1028–1034, November 2016
12. Lameski, J., Jovanov, A., Zdravevski, E., Lameski, P.L., Gievska, S.: Skin lesion segmentation with deep learning. In: IEEE EUROCON 2019–18th International Conference on Smart Technologies. IEEE (2019). https://doi.org/10.1109/EUROCON.2019.8861636
13. Mittal, G., Yagnik, K.B., Garg, M., Krishnan, N.C.: SpotgarBage: smartphone app to detect garbage using deep learning. In: Proceedings of the 2016 ACM International Joint Conference on Pervasive and Ubiquitous Computing, UbiComp 2016, pp. 940–945. ACM, New York (2016)
14. Petrovska, B., Atanasova-Pacemska, T., Corizzo, R., Mignone, P., Lameski, P., Zdravevski, E.: Aerial scene classification through fine-tuning with adaptive learning rates and label smoothing. Appl. Sci. **10**, 5792 (2020)
15. Petrovska, B., Zdravevski, E., Lameski, P., Corizzo, R., Stajduhar, I., Lerga, J.: Deep learning for feature extraction in remote sensing: a case-study of aerial scene classification. Sensors **15**(1), 1 (2020)
16. Prajakta, G., Kalyani, J., Snehal, M.: Smart garbage collection system in residential area. IJRET Int. J. Res. Eng. Technol. **4**(03), 122–124 (2015)
17. Ronneberger, O., Fischer, P., Brox, T.: U-Net: convolutional networks for biomedical image segmentation. In: Navab, N., Hornegger, J., Wells, W.M., Frangi, A.F. (eds.) MICCAI 2015. LNCS, vol. 9351, pp. 234–241. Springer, Cham (2015). https://doi.org/10.1007/978-3-319-24574-4_28
18. Ryan, S., Corizzo, R., Kiringa, I., Japkowicz, N.: Pattern and anomaly localization in complex and dynamic data. In: 2019 18th IEEE International Conference on Machine Learning and Applications (ICMLA), pp. 1756–1763 (2019)
19. Srivastava, N., Hinton, G., Krizhevsky, A., Sutskever, I., Salakhutdinov, R.: Dropout: a simple way to prevent neural networks from overfitting. J. Mach. Learn. Res. **15**(1), 1929–1958 (2014)

Fine – Grained Image Classification Using Transfer Learning and Context Encoding

Marko Markoski and Ana Madevska Bogdanova[✉]

Ss. Cyril and Methodius University in Skopje, Rugjer Boshkovikj, 16, P. O. 393,
1000 Skopje, North Macedonia
marko.markoski@students.finki.ukim.mk,
ana.madevska.bogdanova@finki.ukim.mk

Abstract. Fine - grained classification is a class of problems whose aim is to distinguish between classes of objects that are similar in a great manner. However, this is not an easy task, and difficult tasks require a plethora of tools and creative solutions. In this paper, we present a solution to one fine - grained classification problem, combining deep learning and transfer learning for best end results by using featurization layers of a pretrained neural network as a feature generator, and then feeding those features to our custom neural classifier. Our best results are over 92% overall accuracy.

Keywords: Fine - grained classification · Transfer learning · Deep learning · LSTM · VGG16

1 Introduction

Fine - grained image classification is a very important task for modern computer vision. It aims to separate objects with relatively similar characteristics into different classes as accurately as possible. One type of a fine - grained task is discrimination between species or subspecies, such as bird or monkey species. The second example is the basis of our research. Fine - grained image classification is facing multiple challenges, such as high intra - class variance (objects belonging to the same subclass often exhibit very different visual characteristics, like different poses or viewpoints), low inter - class variance (stemming from inherent similarities of the target objects) and limited fine - grained training data.

Deep learning is a subfield of machine learning which uses deep artificial neural networks. It has been shown to be a very powerful solving tool for many diverse problems, including image classification and recognition, text classification, text generation etc. This applies to fine - grained tasks as well. The major state-of-the-art deep neural networks are all trained on coarse-grained data, which can hurt their performance on fine - grained datasets [1]. However, an advantage of these strong neural networks is that by being able to distinguish among large numbers of classes, they can probably provide high-quality general features that the researchers in computer vision and classification problems can use for classification [z].

© Springer Nature Switzerland AG 2020
V. Dimitrova and I. Dimitrovski (Eds.): ICT Innovations 2020, CCIS 1316, pp. 64–70, 2020.
https://doi.org/10.1007/978-3-030-62098-1_6

The previous discussion brings us to the concept of transfer learning. Transfer learning refers to a technique and a problem in the field of machine learning, which focuses on transferring knowledge gained from solving one problem, to a different related problem. For example, feature extraction layers from proven neural networks can be used as a feature extractor in a different image classification problem.

Long short - term memory (LSTM) networks are recurrent neural networks usually used with sequence data. They have been shown in previous research to be very capable of fusing the complete contextual information of an image [2], which can be very helpful when one is dealing with different but similar classes. In our neural model we utilize neural layers of this type.

In this paper, we present a way of separating fine - grained classes based on deep neural networks and transfer learning. We use the supervised approach, working with a labeled dataset of monkey species. We make use of the featurization layers of a pretrained neural network as a feature generator, and then feed those features to our custom neural classifier, introducing simplified structure compared to some other solutions of the same problem.

The rest of the paper is organized as follows. Section 2 presents the related work to this subject. In Sect. 3 we give a description of the proposed model used in the research. In Sect. 4, we present our results and compare them with other results achieved on the same dataset. In Sect. 5 we summarize our research and propose further use of our approach, while at the end we acknowledge our references.

2 Related Work

In this section, we look at some influential research carried out on our subject of interest.

Weakly supervised approach to fine - grained image classification is an approach which aims to improve the featurization of an image and as a result, the class recognition. This is a very broad approach, but it has been shown in more than one study to be quite successful. In [2], complex featurization and thresholding is performed. Through a complex pipeline of steps, the end features likely contain enough information for class discrimination. This information is fused with long short – term memory neural layers, resulting in high accuracy when tested on multiple datasets [2]. Another inspirational paper [1] is using somewhat less complex featurization, with end features representing important object parts. It fuses coarse - grained and fine - grained prediction in order to obtain the final prediction. This approach is also quite successful on multiple datasets.

Transfer learning is an approach which has also been extensively used in many research papers. The study using LSTM neural layers in [2] makes use of pretrained neural networks during multiple phases of the whole classification process, and then fine - tunes them in order to suit its own purposes. There are also many other researches using fine - tuning an existing neural network as an approach to solving multiple tasks. For example, [3] is using a sub approach called domain adaptive transfer learning, dealing, among other things, with fine - tuning the loss function, so that the probabilistic behavior of the new dataset is taken into consideration as well. It manages to beat other pretraining approaches on many datasets.

Regardless of the used approach, fine - grained image classification remains a big and challenging class of problems.

3   Proposed Solution

Before presenting the actual results, we describe all the integral parts of our final classification model.

3.1   Feature Extraction

A major challenge in deep learning, and machine learning in general, is identifying discriminative features of interest. Deep learning is aiming to replace the feature engineering of mainstream machine learning with feature learning from the input data. This means that deep learning can be a very powerful tool, if utilized properly. But obtaining distinguishing features for accurate classification, especially from fine - grained classes of objects, is a problem. On the other hand, when using the right features, classification proves to be a much easier task, as proved by multiple research papers.

The first part of our solution, the transfer learning part, is an already existing and proven featurizer, the VGG16 neural network [4]. It is a state-of-the-art neural network developed by researchers from the University of Oxford [5] (Fig. 1).

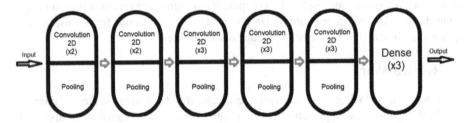

Fig. 1. An overview of the general structure of VGG16

The depicted deep neural model was trained for weeks on GPUs before being tested on a subset of the ImageNet dataset [6, 14, 15]. It contains images belonging to a total of thousand classes, managing to achieve an impressive 92.7% accuracy [4].

We leave out the fully connected (dense) layers and take the convolution - activation - pooling part of VGG16 as a featurizer for our specific problem, thus using an already great amount of knowledge that can be enhanced and adapted to our dataset. It is worth mentioning that we do not fine - tune the pretrained part itself; we merely use it as a feature extractor and then we train our own neural network using the generated features.

3.2   Classification Model for Fine - Grained Image Classification

Having a ready - made featurizer solves one of the greatest challenges of our research. We built a classifier without any convolution whatsoever, because we concluded that the input data is already filtered and pooled enough; instead, our custom model deals with classification of the extracted features.

The proposed custom classification model is motivated by the improvements that LSTM layers had brought to the work presented in [2]. After the recurrent layers, which had been shown to be a very powerful tool for data context encoding in [2], we add classification layers - densely connected layers. In order to simplify our model, we don't use very specialized techniques in the final experimental models with the other hyperparameters.

The detailed implementation of the proposed solution is covered in the next section.

4 Experiments and Results

In this section, we give a detailed description of our experiments and the results.

4.1 Dataset

The dataset used in this research is public and available at the data science platform Kaggle [7]. It is a dataset consisting of 1370 images belonging to one of ten monkey species [8].

The dataset in its official form is divided between train and validation images for each class, the class cardinalities and training - validation splits being relatively uniform. Our experiments were conducted on a unified and shuffled data set. The complete dataset consists of 157 images for the mantled howler, 167 images for the patas monkey, 164 images for the bald uakari, 182 images for the Japanese (snow) macaque, 157 images for the pygmy marmoset, 169 images for the white headed capuchin, 158 images for the silvery marmoset, 170 images for the common squirrel monkey, 160 images for the black - headed night monkey and 158 images for the Nilgiri langur. Their corresponding Latin names are given on Kaggle [8].

4.2 The Implementation

We implemented our neural network using Keras deep learning library (API) [9].

The first step in our final implementation is an appropriate data preprocessing. The images from the dataset come in a variety of shapes. Data shape variation is a problem that is encountered by every image processing task solved by deep learning. In order to resolve this matter, the first preprocessing steps in our model were - computing the average dimensions of all images and then accordingly resizing all the images. The average height of the dataset is 480, and the average width is 394. A single Gaussian pyramid subsampling step on each of the resized images was performed, enabling that the next steps are executing significantly faster, while the lost information is as least as possible. As a next step, we extracted features for each input image using the pretrained part of our classifier. The extracted features have, in the Keras implementation of VGG16, a final shape of $7 \times 6 \times 512$ [10]. The data and its labels were consistently shuffled, in order to make the classifier unbiased to the true physical order of the images.

The structure of the proposed neural classifier is given in Fig. 2. For the recurrent (LSTM) layers to be able to work with the extracted features, a reshaping step is necessary, in the form of a Reshape layer (part of the Keras layer toolkit) [9]. Then we put the recurrent layers on top of the reshaping computational step, and finally two classification layers were used; in the experiments where the recurrent layers return sequences, a

Flatten layer before the final classification is necessary. The classification layer is using the softmax function as an activation function, while the other densely - connected layer and the first recurrent layer are using ReLU as an activation function.

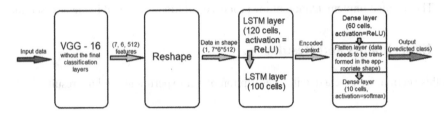

Fig. 2. A general overview of the proposed model

The data set is split into three subsets: training, validation and test set - 80% of the data instances are training entries (1096), and validation and test data entries each make up 10% of the dataset (137 instances each). The model is trained for 6 epochs in the best experiments (empirically obtained as an optimal number) and used a batch size of 3 in most of the experiments. In line with the previously stated simplistic approach to the hyperparameters, we use Adadelta optimizer with the default learning rate (it is given as a string parameter) and ordinary categorical cross - entropy as a loss function.

The evaluation of the proposed model is done by overall classification accuracy, and precision and recall for each class.

4.3 Results

We managed to achieve a **best overall accuracy of 0.927.** This result is comparable with some strong results published on the platform Kaggle [8]. The proposed neural network model was able to determine some of the classes without any false positives and false negatives.

We use class numbering from one to ten, so that each class number from the source dataset is incremented by one [8]. The results are rounded on two decimals.

The detailed per - class precisions obtained on the test dataset are presented in Table 1.

Table 1. Table of precisions

	1	2	3	4	5	6	7	8	9	10
Val.	0.8	0.93	0.89	0.93	0.94	0.89	1.0	1.0	0.92	1.0

In Table 2 we present the per – class recalls obtained on the test dataset.

The results imply that the model learns to distinguish some classes better than the others. Even the worst scores are high, bearing in mind that before the training and testing phases there is a random shuffling step, so we don't really know how much data the model had to learn from.

Table 2. Table of recalls

	1	2	3	4	5	6	7	8	9	10
Val.	0.86	1.0	0.89	0.87	1.0	0.89	0.88	0.94	0.92	0.91

4.4 Comparison with Closely Related Work

There has been research by other authors on this exact dataset as well. The results presented in this paper are comparable to some of the best results published on the Kaggle platform [8].

The best published results on this problem and data set, manages to achieve classification accuracy of 0.99 after only a few epochs of GPU training [11]. It is using the pretrained ResNet34 model, the winner of the 2015 ImageNet competition.

In another research, quite similar in approach to the proposed model in this paper, the end accuracy is more than 0.9 [12]. It is an approach based on transfer learning, and the pretrained part in this case is the Xception neural network [13]. The classifier is simple in this case, containing two densely - connected layers, and those steps are enough for achieving 0.9 accuracy. We improved the accuracy of this approach, and we also improved some of the other metrics. For example, we improved both the precision and recall of the Nilgiri langur, which are both less than 0.9 in [12].

There are other respectable published results, achieving lower or higher accuracy than the model proposed in this paper. Nevertheless, the fine - grained classification is never an easy task, so different solutions can give different distinctions between classes, and the user can choose the model that brings most information for the problem at hand.

5 Conclusion

In this paper we combined deep learning methodology with transfer learning to solve at least one instance of a difficult general problem, segregating similar objects belonging to subclasses of a more general class - a fine-grained image classification problem. We extracted features from the input images using a pretrained neural network, VGG16 [10], and then reshaped them in order to achieve the best end results. The best accuracy of 0.927 was achieved with two LSTM layers and two densely - connected classification layers. Our proposed model contributes to the fine - grained classification task, as different solutions can give improved distinctions between classes compared to other solutions and the user can choose the model that brings most information for a problem at hand.

We believe that the proposed approach is applicable to other fine - grained classification problems. It serves as a further proof of the effectiveness of all used techniques when it comes to fine - grained image classification.

As for future work, we propose using our approach in the medical image classification domain.

References

1. Hu, T., Qi, H., Huang, Q., Lu, Y., et al.: See better before looking closer: weakly supervised data augmentation network for fine-grained visual classification. https://arxiv.org/pdf/1901.09891.pdf. Accessed 10 Apr 2020
2. Ge, H., Lin, X., Yu, Y., et al.: Weakly supervised complementary parts models for fine-grained image classification from the bottom up. https://arxiv.org/abs/1903.02827. Accessed 10 Apr 2020
3. Ngiam, J., et al.: Domain adaptive transfer learning with specialist models. https://arxiv.org/pdf/1811.07056.pdf. Accessed 15 Apr 2020
4. VGG16 – Convolutional Network for Classification and Detection. https://neurohive.io/en/popular-networks/vgg16/. Accessed 23 Aug 2020
5. Simonyan, K., Zisserman, A.: Very deep convolutional networks for large-scale image recognition. https://arxiv.org/abs/1409.1556v5. Accessed 23 Aug 2020
6. ImageNet Dataset. http://image-net.org/. Accessed 23 Aug 2020
7. Kaggle: A Platform For Machine Learning and Data Science. https://www.kaggle.com/. Accessed 2020/8/23
8. Monkey Species Dataset. https://www.kaggle.com/slothkong/10-monkey-species. Accessed 20 Aug 2020
9. Keras: A Deep Learning API. https://keras.io/. Accessed 20 Aug 2020
10. A Keras Implementation of the VGG16 NN. https://github.com/keras-team/keras-applications/blob/master/keras_applications/vgg16.py. Accessed 20 Aug 2020
11. Fast.ai With ResNet34 (99% Accuracy). https://www.kaggle.com/pinkquek/fast-ai-with-resnet34-99-accuracy. Accessed 20 Aug 2020
12. Monkey Classifier CNN (Xception ~ 0.90 Acc. https://www.kaggle.com/crawford/monkey-classifier-cnn-xception-0-90-acc. Accessed 20 Aug 2020
13. Chollet, F.: Xception: deep learning with depthwise separable convolutions. In: CVPR (2017)
14. Russakovsky, O., et al.: Imagenet large scale visual recognition challenge. Int. J. Comput. Vision 115(3), 211–252 (2015)
15. Krizhevsky, A., Sutskever, I., Hinton, G.E.: ImageNet classification with deep convolutional neural networks. In: Advances in Neural Information Processing Systems, pp. 1097–1105 (2012)

Machine Learning and Natural Language Processing: Review of Models and Optimization Problems

Emiliano Mankolli[1]([⊠]) and Vassil Guliashki[2]

[1] Faculty of Information Technology, "Ritech Group of Companies", Machine Learning Team, "Aleksandër Moisiu" University of Durrës, Durrës, Albania
emiliano_fshn@hotmail.com
[2] Institute of Information and Communication Technologies, Acad. G. Bonchev Street 2, 1113 Sofia, Bulgaria
vggul@yahoo.com

Abstract. The purpose of this paper is to consider one of the most important modern technologies, namely: Natural Language Processing (NLP) and the machine learning algorithms related to it. The aim of the authors is to present the machine learning models in this interdisciplinary scientific field, which represents an intersection point of computer science, artificial intelligence, and linguistics. The machine learning techniques are classified and the corresponding models are briefly discussed. Different optimization approaches and problems for machine learning are considered. In this regard, some conclusions are drawn about the development trends in the area and the directions for future research.

Keywords: Natural Language Processing · Machine learning · Linguistic rules · Optimization methods

1 Introduction

Natural Language Processing (NLP) or computational linguistics is one of the most important modern technologies. It is interdisciplinary in its essence and is based on linguistics, computer science, and artificial intelligence. The machine learning techniques are connected with NLP and they have to "understand" the natural language and to perform different complex tasks like Language Translation, Question Answering, etc. Accurate understanding the meaning of the human language message is extremely difficult due to the following features: 1) The human language is a complex, discrete, symbolic signal system, providing high reliability of the signal. 2) Definite symbols in a specific language can be coded as a signal for communication in various ways, such as writing, sounds, gestures, images, etc. The main inseparable elements of language are words. 3) Some of the words in the language have multiple meanings (polysemy). Often their different meanings are entirely opposite in nature (auto-antonyms) [1]. There are sets of words having the same meaning (synonyms). Also, some words behave differently when used as nouns or verbs [1]. In addition, there are idiomatic combinations

© Springer Nature Switzerland AG 2020
V. Dimitrova and I. Dimitrovski (Eds.): ICT Innovations 2020, CCIS 1316, pp. 71–86, 2020.
https://doi.org/10.1007/978-3-030-62098-1_7

of words, that are characterized by a specific meaning. 4) Different human languages are structured differently by specific grammatical rules. In general, the texts can contain *lexical* (multiple meanings of one word), *syntactic* (connected with morphology [2]), *semantic* (sentence having multiple meanings) [3], and *anaphoric ambiguity* (phrase or word, mentioned previously, but with a different meaning).

For this reason, there is a "high level of complexity in presenting, learning, and using language/situational/contextual/word/visual knowledge" based on human language [4]. From another side "computers interact with humans in programming languages which are unambiguous, precise and often structured" [1].

Various tasks of NLP are performed in order to clarify/reduce the ambiguity in the verbal or textual linguistic representation of knowledge. With this aim, powerful analysis tools such as statistical techniques and machine learning are used. Machine learning involves the use of optimization methods, which in turn have a significant impact on the development of this area itself. The purpose of this paper is to review the machine learning techniques in natural language processing, classified according to the models used, and then to make a brief overview of the main approaches and optimization methods used in this scientific field.

A survey on the state-of-the-art machine learning models for NLP is presented in [3]. Optimization methods from a machine learning perspective are surveyed in [5].

The content of the paper is organized as follows: The basic machine learning techniques are presented in Sect. 2. The most important machine learning models are considered in Sect. 3. Section 4 is devoted to the optimization problems, which have to be formulated in the machine learning in connection with NLP. Conclusions are drawn in Sect. 5.

2 Machine Learning Techniques

There are two main approaches used in the natural language processing, namely: 1) language processing based on linguistic rules (LR) and 2) machine learning (ML), in which algorithms for statistical machine learning are applied [3, 6]. From the viewpoint of statistical methods for NLP, there are two main categories: clustering techniques and machine learning algorithms. ML attempts to solve the knowledge extraction problem by means of a classification process, whereas the clustering is based on a similarity measure. ML algorithms are broadly used not limited only in the NLP area but in many other domains.

In the LR approach, the developer uses a linguistic engine that has knowledge about the syntax, semantics, and morphology of a language [6] and then adds program rules that search for key semantic concepts determining the meaning of a particular verbal expression. These rules are constructed manually by grammarians or language experts to perform various specific NLP tasks.

Machine learning (ML) applies neural networks and intelligent modules that are able to learn from historical data. In ML, the rules for language processing are derived from the training data. Statistical machine learning algorithms use historical data and synthesize textual characteristics, which are used for classification or prediction. Modern research in this field uses ML models. Real specific weight values are attached to the introduced characteristics in the models, and thus probabilistic solutions are then generated. The advantage of the models is that they are able to represent "the relation quality in different dimensions" [3]. The rule extraction in machine learning can be done by solving an optimization task. Two genetic-algorithm-based approaches for finding non-dominated rule sets are described and compared in [7]. The linguistic rule extraction is formulated as a three-objective combinatorial optimization problem. Statistical and probabilistic methods are used to perform NLP tasks.

Modern techniques that are widely used to perform the main tasks of NLP are machine learning techniques. ML techniques can generally be divided into three categories [3]: (1) supervised machine learning, (2) semi-supervised machine learning, and (3) unsupervised machine learning [2].

The most popular technique used to perform NLP tasks is supervised machine learning, which models are based on automatic extraction of rules from training data [3]. Supervised ML can be classified as (i) sequential and (ii) non-sequential.

Sequential supervised ML techniques are Deep Learning (DL), Conditional Random Fields (CRF), k-Nearest Neighbors (k-NN), Hidden Markov Model (HMM), and Maximum Entropy (MaxEnt). Non-sequential supervised ML techniques include Naive Bayes, Decision Trees (DT), and Support Vector Machines (SVM).

In semi-supervised machine learning, a small degree of supervision is used. An example of this technique is bootstrapping.

Unsupervised machine learning uses models, which are not trained. Search for intra-similarity and inter-similarity between objects is performed to solve ML tasks (see [3]). The most popular approaches in this ML category are clustering and vector quantization (see [8]).

3 Most Important Machine Learning Models

3.1 Supervised ML Models

A. Sequential Models

- *Deep learning (DL)*

Deep learning is a branch in machine learning based on the use of artificial neural networks. It has become very popular relative recently and can perform multiple levels of abstraction learning. Neural networks have been introduced in 1958 [9] but they are more popular only since 2012 [10] with the use of large labeled data sets and of Graphics Processing Units (GPU) [11]. "Deep learning" means that a layered model of inputs, known as neural network uses several layers [12] to learn multiple levels of representation or of increasing complexity (abstraction) [13]. Each next layer uses the output of the previous layer and feeds its result to the successive layer.

DL networks are used for natural language processing [14] and speech processing [15]. They can also perform tasks on sentiment analysis, parsing, Name Entity Recognition (NER), and other areas of NLP. Chinese word segmentation and Part Of Speech (POS) tasks are performed using DL approach in [16]. A DL neural network architecture for word level and character level representations, as well as for POS tagging is proposed in [17]. Word segmentation by means of a perceptron based learning algorithm using character based classification system is performed in [18]. A deep neural network system for the information extraction task is proposed in [19]. The Arabic NER task in three phases is performed in [20]. Results show that the neural network outperforms the decision tree approach. In [21] an evaluation platform is developed and used to improve the efficiency of sentiment analysis models in the finance area. Different word and sentence embeddings in combination with standard machine learning and deep-learning classifiers for financial texts sentiment extraction are evaluated in [22]. The evaluation shows that the BiGRU + Attention architecture with word embedding as features, give the best score.

- *Conditional Random Field (CRF) Model*

Conditional random field is introduced in [23]. This is a statistical model used for pattern recognition and machine learning applying structured prediction. In the NLP area, CRF models are the most adaptable models for many tasks [1]. The CRF model is based on conditional distribution instead of joint distribution of a set of response instances [24].

CRF is defined in [23] on X - random variable over data sequences to be labeled and Y - random variable over corresponding label sequences. All components Y_i of Y are assumed to range over a label alphabet, which is finite. The random variables X and Y are jointly distributed. A conditional model $p(Y|X)$ from paired observation and label sequences is constructed. The marginal $p(X)$ is not explicitly modeled.

Let $G = (V, E)$ be a graph such that $Y = (Y_v)_{v \in V}$, so that Y is indexed by the vertices of G. Then (X, Y) is a conditional random field when the random variables Y_v, conditioned on X, obey the Markov property with respect to the graph:

$$p(Y_v|X, Y_w, w \neq v) = p(Y_v|X, Y_w, w \sim v), \tag{1}$$

where $w \sim v$ means that w and v are neighbors in G.

CRF is considered as a random field globally conditioned on the observation X and the graph G is assumed to be fixed. The idea in the CRF model is to maximize a log-likelihood objective function finding the label y, which maximizes the conditional likelihood $p(y \mid x)$ for a sequence x, $(X = x, Y = y)$. Generally speaking, CRF models are feature-based models and can operate on binary or real valued features/variables.

The simplest example for modeling sequences is when G is a simple chain or line: $G = (V = \{1, 2, \dots m\}, E = \{(i, i + 1)\})$. X may also have a natural graph structure, but here the sequences $X = (X_1, X_2, \dots, X_n)$ and $Y = (Y_1, Y_2, \dots, Y_n)$ will be considered.

In case the graph $G = (V, E)$ of Y is a tree (where the chain is the simplest example), the edges and the vertices are its cliques. By the fundamental theorem of random fields [25] the joint distribution over label sequence Y given X is presented as:

$$p(y|x) = \frac{1}{Z(x)} \exp \left(\sum_{e \in E, k} \lambda_k f_k(e, y|_e, x) + \sum_{v \in V, k} \mu_k g_k(v, y|_v, x) \right) \tag{2}$$

where x is a data sequence, y is a label sequence, and $y|_S$ is the set of components of y associated with the vertices in sub-graph S. $Z(x)$ is a normalization factor, f_k and g_k are k-th feature functions, and λ_k and μ_k are the correspondent weights for them.

Conditional random fields approach for the NER task is presented in [26]. A method of NER is proposed, which combines the hybridization of a k-Nearest Neighbors (k-NN) classifier with a conventional linear CRF model. Bootstrapping technique for Arabic named entity recognition applying a CRF model is considered in [27]. The proposed approach is used for a named entity recognition problem in Arabic language. The NER problem in the Chinese language in connection with a CRF model using a pool-based active learning algorithm is considered in [28].

Models for language processing based on the CRF model for Part of Speech (POS) tagging are proposed in [29–31].

- Hidden Markov Model (HMM)

The Hidden Markov Model (HMM) is connected with both observed events (like words that can be seen in the input) and hidden events (like part-of-speech tags) that are perceived as causal factors in a probabilistic model [32]. HMM also performs a sequence classification or sequence labeling. In the case of a sequence classifier, it firstly identifies a class label for each token and then assigns the corresponding label to each token or word of the given input sequence. HMM has great success in textual classification – POS tagging, for example name recognition tasks [33]. The aim is to recognize person names, location names, brand names, designation, date, time, abbreviations, number, etc. and to classify them into given predefined different categories [34]. In such a task there are assigned labels to the words in context (a single label to a word). The states of the HMM model are organized into regions, one region for each desired class and one for Not-A-Name. There are also two special states: the state Start-Of-Sentence, and the state End-Of-Sentence. A statistical bigram language model is used to compute the probability of a sequence of words occurring within each region (name-class (NC)). The model employs a Markov chain, where the probability of each word is based on the previous word. Every word is represented by a state s_i in the bigram model. A probability is associated with the transition from the current to the next word. There are two assumptions in HMM [35]: 1) Markov assumption:

$$P(s_i|s_1 \ldots s_{i-1} = P(s_i|s_{i-1}) \tag{3}$$

and 2) Output independent assumption:

$$(o_i|s_1 \ldots s_i, \ldots, s_T, o_1, \ldots, o_i, \ldots o_T) = P(o_i|s_i) \tag{4}$$

where the probability of the resultant observation O_i is calculated using the probability only of the state s_i generating the observation and does not depend on the other states around it.

The HMM model uses a vector of initial probabilities, a matrix of transition probabilities and a matrix of observation probabilities. In this model, the probability of a sequence of words w_1 through w_n is determined by:

$$p = \prod_{i=1}^{n} p_i(w_i|w_{i-1})$$

(5)

The probability of w_1 is determined using a special *"begin"* word. For the statistical bigram model, it holds that the number of states in each of the name-class regions is equal to the vocabulary size $|V|$.

HMM is a probabilistic generative model because it generates a sequence of words and labels. In the name recognition task, the most likely sequence of name-classes (*NC*) given a sequence of words (*W*) is found:

$$\max P(NC|W) = \frac{P(W, NC)}{P(W)},$$

(6)

according to the Bayes' Rule. The denominator in (6) is the unconditioned probability of the word sequence and is constant for any given sentence. Hence, the maximization only of the numerator is necessary.

HMM for the NER task is proposed in [36]. It operates on sentence wise data and assigns corresponding NE tag to each word in the sentence. NER for Hindi, Marathi, and Urdu languages applying HMM with high levels of accuracy is considered in [37]. A rule-based method in combination with HMM is used for POS tagging in [38]. Sentence boundary detection (SBD) tasks are considered in [39, 40]. The word segmentation problem is solved applying HMM in [41, 42].

- Maximum entropy model (MaxEnt)

The maximum entropy model is a general-purpose model for making predictions or inferences from incomplete information. It is used for sequential data classification.

The task of this model is performed in three steps: 1) extracting relevant features from the given input sequence, 2) performing linear combination of the extracted features, and 3) taking exponent of resulted sum [35]. The probability distribution of a certain class 'x' given the observation 'o' is given as:

$$P(x|o) = \frac{1}{Z} \exp\left(\sum_{i=1}^{n} w_i f_i\right),$$

(7)

where Z is a normalization function, and $exp = e^x$.

In [43] is proposed a NER system based on maximum entropy (MaxEnt). A POS tagging system using MaxEnt model is developed in [44].

- k-Nearest Neighbors (k-NN)

The k-Nearest Neighbors (k-NN) algorithm [45–47] is a simple instance-based or a lazy supervised machine learning algorithm. It can be used for regression analysis [45, 48].

k-NN is a non-probabilistic and non-parametric model [47], in which no explicit training step is required [49]. This model is very suitable for classifications in case that there is no prior knowledge about the distribution of data. The classification is based on distance metrics. The key idea of the k-NN algorithm is first to select k-nearest neighbors for each test sample and then to use them to predict this test sample. The complexity increases with the dimensionality increasing. For this reason, dimensionality reduction techniques [50] have to be applied before using k-NN.

B. **Non-sequential Models**

- *Naive Bayes*

The Naive Bayes classifier is a supervised machine learning algorithm (a data set which has been labeled) based on the popular Bayes theorem of probability. The likelihood that a given specimen has a place with a specific class is dependent upon Bayes hypothesis. The term "Naive" is used on the base of the strong assumption, that all the input features are independent of each other and no correlation exists between them. This classifier is used for binary as well as multi-class statistical classification problems especially in the field of document classification [51, 52]. The Naive Bayes classifier needs a little measure of preparing set to determine the parameters for characterization [53].

- *Decision trees (DT)*

Decision trees serve for classification/prediction in ML. DT model is a tree, for which each node is a decision, and each leaf is an output class. The node can have more than two children, but most algorithms use only binary trees. DT were first applied to language modeling in [54] to estimate the probability of spoken words. A single node is the starting point followed by binary questions that are asked as a method to arbitrarily partition the space of histories. As space is partitioned, "leaves" are formed, and training data is used to calculate the conditional probability of $P(w|h)$ for the next element. As the traversal continues, the questioning becomes more informative by the use of information theoretic metrics. Such metrics include Kolomogorov Complexity, entropy, relative entropy, etc. [55]. In [56] are proposed: 1) a tree-pruning method that uses the development set to delete nodes from over-fitted models, and 2) a result-caching method. As a result, the developed algorithm is 1 to 3 orders of magnitude faster than a naive implementation and performs accurately on data sets, which refer to a) letter-to-sound (LTS); b) syllabification; c) part-of-speech tagging; d) text classification, and e) tokenization.

- *Support vector machines (SVM)*

SVM is used mainly for binary classification of linear, as well as nonlinear data. This model is used to separate data across a decision boundary (hyperplane) determined by only a small subset of the data (feature vectors). In the field of natural language processing, SVMs are applied to many tasks as, for example: content arrangement, POS, NER, segmentation, etc. The classification rule for the separating hyperplane can be written as:

$$f(x, w, b) = W.X + b = 0 \qquad (8)$$

where $W = w_1, w_2, w_3, \ldots, w_n$ is a weight vector, n is the number of attributes; x is the example to be classified and b is a scalar, often referred to as bias. Equation (8) can be written as:

$$w_0 + w_1 x_1 + \ldots + w_n x_n = 0 \qquad (9)$$

where b is represented as an additional weight, w_0, and x_1 and x_2 are the values of attributes A_1 and A_2, for X respectively.

A combination of SVM and CRF is used in [57] for Bengali named entity recognition. In [58] a POS tagger for Malayalam language is created using SVM.

3.2 Semi-supervised ML Models

- *Bootstrapping*

In [59] is presented the Dual, Iterative Pattern Relation Expansion (DIPRE) system, which applies bootstrapping for extracting a relation of books − (author, title) from the Web. This system represents the occurrences of seeds as three contexts of strings: words before the first entity (BEF), words between the two entities (BET), and words after the second entity (AFT). DIPRE generates extraction patterns by grouping contexts based on string matching and controls semantic drift by limiting the number of instances a pattern can extract.

In [60] is considered the Snowball system, which is inspired on DIPRE's method of collecting three contexts for each occurrence but computing a Term Frequency - Inverse Document Frequency (TF-IDF) representation for each context. The seed con-texts are clustered through an algorithm based on the cosine similarity between contexts using vector multiplication between corresponding sentence score vectors [61].

3.3 Unsupervised ML Models

- *Clustering*

The clustering model [62, 63] is used to find a structure or pattern in a collection of unlabeled data sets. The clustering algorithm groups the data of a given data set into K clusters. The data points within each cluster are similar to each other and the data points in different clusters are dissimilar. Analog to the k-NN algorithm, a similarity metric or distance metric, such as Euclidean, Mahalanobis, cosine, Minkowski, etc. is used. The Euclidean distance metric is used very often, but in [64] it is shown that this metric is unable to guarantee the quality of the clustering. The K-means algorithm is one of the simplest clustering algorithms. It separates the data points into K groups of equal variances, minimizing the within-cluster sum-of-squares or the inertia.

- *Vector quantization*

The vector quantization model [65–67] organizes data in vectors and represents them by their centroids. Usually, the K-means clustering algorithm is used to train the quantizer. Codewords are formed by the centroids. All codewords are stored in a Codebook. Vector quantization is a method for loss compression. It is used in many coding applications. The errors in the compressed data are inversely proportional to density. The Vector quantization model is used in speech coding [66, 68], audio compression [69], and many other applications.

4 Optimization Problems in ML

Usually in machine learning algorithms, an optimization problem is formulated and the extremum of an objective function is searched. The first step in a ML method is the determination of the model and the choice of a reasonable objective function. Then a suitable optimization method is used to solve the corresponding optimization problem.

According to the models considered in Sect. 3, ML optimization problems are defined for supervised learning, semi-supervised learning, and unsupervised learning. In supervised learning, there are formulated sentence classification problems [70, 71], and regression problems [72]. Unsupervised learning is divided into clustering and dimension reduction [73].

4.1 Supervised Learning Optimization Problems

The aim here is to formulate an optimal mapping function $f(x)$, and then to minimize the loss function of the training samples:

$$\min_{\theta} \frac{1}{N} \sum_{i=1}^{N} L(y^i, f(x^i, \theta)) \tag{10}$$

where N is the number of training samples, x^i is the feature vector of the i-th sample, y^i is the corresponding label, θ is the corresponding parameter of the mapping function, and L is the loss function.

Different kinds of loss functions can be used, such as the square of Euclidean distance, contrast loss, hinge loss, cross- entropy, information gain, etc. Another optimization problem is the structured risk minimization, which is solved usually by the support vector machine method. Regularization items are included in the objective function to reduce/avoid the overfitting. Usually, they are in form of L_2 norm:

$$\min_{\theta} \frac{1}{N} \sum_{i=1}^{N} L(y^i, f(x^i, \theta)) + \lambda \|\theta\|_2^2 \tag{11}$$

where the compromise parameter λ can be determined by means of cross-validation.

4.2 Semi-supervised Learning Optimization Problems

Semi-supervised learning (SSL) is the technique between supervised and unsupervised learning. It incorporates labeled data and unlabeled data during the training process and can be used for the following tasks: classification [74], clustering [75], regression [76], and dimensionality reduction [77]. There are different kinds of semi-supervised learning methods: generative models, semi-supervised support vector machines (SSSVM) [78], self-training, graph-based methods, multi-learning method, and others.

For example, SSSVM is a learning model that can deal with binary classification problems. In this case, only a part of the training set is labeled. Let D^l be labeled data: $D^l = \{\{x^1, y^1\}, \{x^2, y^2\},...,\{x^l, y^l\}\}$, and D^u be unlabeled data: $D^u = \{x^{l+1}, x^l+^2, ..., x^N\}$ with $N = l + u$. An additional constraint on the unlabeled data is added to the original objective of SVM with slack variables β^i in order to use the information of unlabeled data. Specifically, define e^j as the misclassification error of the unlabeled instance if its true label is positive and z^j as the misclassification error of the unlabeled instance if its true label is negative. The SSSVM problem can be formulated as:

$$\min\|\omega\| + \gamma \left[\sum_{i=1}^{l} \beta^i + \sum_{j=l+1}^{N} \min(\alpha^i, z^j) \right],$$

subject to

$$y^i(\mathbf{w} \cdot x^i + b) + \beta^i \geq 1, \ \beta^i \geq 0, \ i = 1, \ldots, l,$$
$$\mathbf{w} \cdot x^j + b + a^j \geq 1, \ a^j \geq 0, \ j = l+1, \ldots, N,$$
$$-(\mathbf{w} \cdot x^j + b) + z^j \geq 1, \ z^j \geq 0; \tag{12}$$

where γ is a penalty coefficient. The optimization problem (12) is an NP-hard mixed-integer problem (a mixed integer problem at least as hard as any problem solvable only by a Nondeterministic Polynomial time algorithm) [79]. There are different methods [80], which can be used to solve this problem, such as convex relaxation methods [81] and branch and bound methods [82].

4.3 Unsupervised Learning Optimization Problems

In the clustering algorithms [83], a group of samples is divided into multiple clusters, where the differences between the samples in the same cluster are as small as possible. The samples in different clusters should be as different as possible. The optimization problem for the k-means clustering is minimizing the loss function:

$$\min_{S} \sum_{k=1}^{K} \sum \|x - c_k\|_2^2 \tag{13}$$

where K is the number of clusters, x is the feature vector of the samples, c_k is the center of cluster k, and S_k is the sample set of cluster k. The minimization of this objective function has to make te sum of variances of all clusters as small as possible.

In the dimensionality reduction algorithm, the data are projected into a low-dimensional space and the original information from them should be retained as much as possible after that. The principal component analysis (PCA) [84] algorithm for dimensionality reduction minimizes the reconstruction error:

$$\min \sum_{i=1}^{N} \left\| \bar{x}^i - x^i \right\|_2^2, \text{ where } \bar{x}^i = \sum_{j=1}^{d'} z_j^i e_j \text{ , } d >> d'. \tag{14}$$

where N represents the number of samples, x^i is a d-dimensional vector, $\bar{x}^i$ is the reconstruction of x^i. The projection of x^i in d'-dimensional space is $z^i = \{z_1^i, \ldots, z_{d'}^i\}$, and e_j is the standard orthogonal basis in the d'-dimensional coordinates.

In the probabilistic models often the maximum of a probability density function of $p(x)$ should be found. It represents the logarithmic likelihood function of the training samples:

$$\max \sum_{i=1}^{N} \ln p(x^i, \theta) \tag{15}$$

The parameter θ has the effect of reducing the overfitting. Some prior distributions are often assumed on this parameter in the framework of Bayesian methods.

5 Conclusions

The comparison of ML models and techniques performance is closely related to the specific fields one applies them to.

HMM holds strict independence assumptions, making other models as CRF easier to accommodate any context information. In presence of enough training data, the continuous HMM usually outperforms in accuracy the semi-continuous HMM and the discrete HMM. From an optimization point of view, CRF calculates the normalization probability in the global scope, while in the case of MaxEnt local variance normalization is used. SVM and k-NN are classified as memory-intensive models, resulting in low performance for high dimensional data.

In practice, Decision Tree ensembles are almost always better to use. They are robust to outliers, scalable, and able to naturally model non-linear decision boundaries thanks to their hierarchical structure. They do not require an extensive feature processing but can offer a good overview of the task and dataset itself and can guide the constructing feature templates for other classifiers. In some cases, neural networks outperform the Decision Tree approach, as mentioned in Subsect. 3.1.

In the last years, different pre-trained language models such as ULMFiT, CoVe, ELMo, OpenAI GPT, BERT, OpenAI GPT-2, XLNet, RoBERTa, ALBERT have been developed, which made practical applications of NLP significantly cheaper, faster, and easier, because they allow to pre-train an NLP model on one large dataset and then quickly fine-tune it to adapt to other NLP tasks [85].

Deep Learning in combination with simple reasoning has been used for a long period for speech recognition, but obviously new paradigms are needed to replace rule-based manipulation of symbolic expressions by operations on large vectors. Systems using recurrent neural networks will become much better in understanding sentences or whole documents when they learn strategies for selectively attending to one part at a time. Nowadays Deep Learning algorithms are not suitable for general-purpose executions, as they require a large amount of data.

It can be expected that unsupervised machine learning will become very important in the future. Principally, animal and human learning are unsupervised. The structure of the world is normally discovered by observing it, and not abstractly by receiving the names of all objects. In some areas, the popularity of unsupervised ML methods is growing. For example, a promising Local Aggregation approach to object detection and recognition with unsupervised learning is presented in [86].

Linguistics and human knowledge could improve further the performance of NLP models and systems. Linguistics could support deep learning by improving the interpretability of the data-driven approach. Novel approaches to unsupervised and semi-supervised machine translation have been developed and the quality of the neural machine translation is in progress.

References

1. Agarwal, S.: Word to Vectors — Natural Language Processing (2017). https://towardsdatas cience.com/word-to-vectors-natural-language-processing-b253dd0b0817. Accessed 04 May 2020
2. Daud, A., Khan, W., Che, D.: Urdu language processing: a survey. Artif. Intell. Rev. **47**(3), 279–311 (2016). https://doi.org/10.1007/s10462-016-9482-x
3. Khan, W., Daud, A., Nasir, J.A., Amjad, T.: A survey on the state-of-the-art machine learning models in the context of NLP. Kuwait J. Sci. **43**(4), 95–113 (2016)
4. Le, J.: The 7 NLP techniques that will change how you communicate in the future (Part I) (2018). https://heartbeat.fritz.ai/the-7-nlp-techniques-that-will-change-how-you-commun icate-in-the-future-part-i-f0114b2f0497. Accessed 04 May 2020
5. Sun, A., Cao, Z., Zhu, H., Zhao, J.: A survey of optimization methods from a machine learning perspective. IEEE Trans. Cybern. 1–14 (2019)
6. Goebel, T.: Machine learning or linguistic rules: two approaches to building a chatbot (2017). https://www.cmswire.com/digital-experience/machine-learning-or-linguistic-rules-two-approaches-to-building-a-chatbot/. Accessed 10 May 2020
7. Ishibuchi, H., Nakashima, T., Murata, T.: Multiobjective optimization in linguistic rule extraction from numerical data. In: Zitzler, E., Thiele, L., Deb, K., Coello Coello, C.A., Corne, D. (eds.) EMO 2001. LNCS, vol. 1993, pp. 588–602. Springer, Heidelberg (2001). https://doi.org/10.1007/3-540-44719-9_41
8. Shanthamallu, U.S., Spanias, A., Tepedelenlioglu, C., Stanley M.: A brief survey of machine learning methods and their sensor and IoT applications. In: Proceedings of the 2017 8th International Conference on Information, Intelligence, Systems and Applications, IISA, Larnaca, Cyprus, (2017)
9. Rosenblatt, F.: The perceptron: a probabilistic model for information storage in the brain. Psychol. Rev. **65**, 386–408 (1958)
10. Krizhevsky, A., et al.: Image Net classification with deep convolutional NN. Advances in Neural Information Processing Systems, Vol. **25**. pp. 1090–1098 (2012)
11. Sattigeri, P., Thiagarajan, J.J., Ramamurthy, K.N., Spanias, A.: Implementation of a fast image coding and retrieval system using a GPU. In: 2012 IEEEESPA, Las Vegas, NV, pp. 5–8 (2012)

12. Schmidhuber, J.: Deep learning in neural networks: an overview. Neural Netw. **61**, 85–117 (2015)
13. Deng, L., Yu, D.: Deep learning. Sig. Process **7**, 3–4 (2014)
14. Mikolov, T., Sutskever, I., Chen, K., Corrado, G.S., Dean, J.: Distributed representations of words and phrases and their compositionality. In: Advances in NIPS, October 2013
15. Yu, D., Deng, L.: Automatic Speech Recognition. SCT. Springer, London (2015). https://doi.org/10.1007/978-1-4471-5779-3
16. Zheng, X., Chen, H., Xu, T.: Deep learning for Chinese word segmentation and pos tagging. In: Proceedings of the Conference on EMNLP-ACL-2013, pp. 647–657 (2013)
17. Santos, C.D., Zadrozny, B.: Learning character-level representations for part-of-speech tagging. In: Proceedings of the 31th International Conference on Machine Learning, pp. 1818–1826 (2014)
18. Li, Y., Miao, C., Bontcheva, K., Cunningham, H.: Perceptron learning for Chinese word segmentation. In: Proceedings of Fourth Sighan Workshop on Chinese Language Processing (Sighan-05), pp. 154–157 (2005)
19. Qi, Y., Das, S.G., Collobert, R., Weston, J.: Deep learning for character-based information extraction. In: de Rijke, M., et al. (eds.) ECIR 2014. LNCS, vol. 8416, pp. 668–674. Springer, Cham (2014). https://doi.org/10.1007/978-3-319-06028-6_74
20. Mohammed, N.F., Omar, N.: Arabic named entity recognition using artificial neural network. Journal of Computer Science **8**, 1285–1293 (2012)
21. Mishev, K., Gjorgjevikj, A., Vodenska, I., Chitkushev, L.T., Trajanov, D.: Evaluation of sentiment analysis in finance: from lexicons to transformers. IEEE Access **8**, 131662–131682 (2020)
22. Mishev, K., et al.: Performance evaluation of word and sentence embeddings for finance headlines sentiment analysis. In: Gievska, S., Madjarov, G. (eds.) ICT Innovations 2019. CCIS, vol. 1110, pp. 161–172. Springer, Cham (2019). https://doi.org/10.1007/978-3-030-33110-8_14
23. Lafferty, J., McCallum, A., Pereira, F.C.: Conditional random fields: Probabilistic models for segmenting and labeling sequence data. In: Proceedings of the Eighteenth International Conference on Machine Learning, (ICML), pp. 282–289 (2001)
24. Yang, E., Ravikumar, P., Allen, G.I., Liu, Z.: Conditional random fields via univariate exponential families. In: Proceedings of Advances in Neural Information Processing Systems (NIPS), vol. 26 (2013)
25. Hammersley, J.M., Clifford, P.: Markov fields on finite graphs and lattices, Computer Science (1971)
26. Liu, X., Wei, F., Zhang, S., Zhou, M.: Named entity recognition for tweets. ACM Trans. Intell. Syst. Technol. (TIST) **4**(1), 1524–1534 (2013)
27. Abdelrahman, S., Elarnaoty, M., Magdy, M., Fahmy, A.: Integrated machine learning techniques for Arabic named entity recognition. IJCSI **7**(4), 27–36 (2010)
28. Yao, L., Sun, C., Li, S., Wang, X., Wang, X.: CRF-based active learning for Chinese named entity recognition. In: Proceedings of the 2009 IEEE International Conference on Systems, Man, and Cybernetics, pp. 1557–1561 (2009)
29. Ammar, W., Dyer, C., Smith, N.A.: Conditional random field auto encoders for unsupervised structured prediction. In: Proceedings of the Advances in Neural Information Processing Systems (NIPS-2014), vol. 26, pp. 1–9 (2014)
30. Pandian, S.L., Geetha, T.: CRF models for Tamil part of speech tagging and chunking. *Proceedings of Int. Conf. on Computer Proc. of Oriental Languages.* pp. 11–22, (2009)
31. Patel, C., Gali, K.: Part-of-speech tagging for Gujarati using conditional random fields. In: Proceedings of the IJCNLP-08 Workshop on NLP for Less Privileged Languages, pp. 117–122 (2008)

32. Jurafsky, D., Martin, J.H.: Speech and Language Processing (3rd ed. draft) (2019). https:// web.stanford.edu/~jurafsky/slp3/. Accessed 13 May 2020
33. Bikel, D.M., Schwartz, R.L., Weischedel, R.M.: An algorithm that learns what's in a name. Mach. Learn. **34**, 211–231 (1999)
34. Singh, U., Goyal, V., Lehal, G.S.: Named entity recognition system for Urdu. In: Proceedings of COLING 2012: Technical Papers, pp. 2507–2518 (2012)
35. Jurafsky, D., James, H.: Speech and Language Processing an Introduction to Natural Language Processing, Computational Linguistics, and Speech. Prentice Hall, Upper Saddle River (2000)
36. Morwal, S., Chopra, D.: NERHMM: a tool for named entity recognition based on hidden Markov model. Int. J. Nat. Lang. Comput. (IJNLC) **2**, 43–49 (2013)
37. Morwal, S., Jahan, N.: Named entity recognition using hidden Markov model (HMM): an experimental result on Hindi, Urdu and Marathi languages. Int. J. Adv. Res. Comput. Sci. Softw. Eng. **3**(4), 671–675 (2013)
38. Youzhi, Z.: Research and implementation of part-of-speech tagging based on hidden Markov model. In: Proceedings of Asia-Pacific Conference on Computational Intelligence and Industrial Applications (PACIIA), pp. 26–29 (2009)
39. Kolar, J., Liu, Y.: Automatic sentence boundary detection in conversational speech: a cross-lingual evaluation on English and Czech. In: Proceedings of International Conference on Acoustics Speech and Signal Processing (ICASSP), pp. 5258–5261 (2010)
40. Rehman, Z., Anwar, W.: A hybrid approach for Urdu sentence boundary disambiguation. Int. Arab J. Inf. Tech. (IAJIT) **9**(3), 250–255 (2012)
41. Gouda, A.M., Rashwan, M.: Segmentation of connected Arabic characters using hidden Markov models. In: Proceedings of the IEEE International Conference on Computational Intelligence for Measurement Systems and Applications (CIMSA), pp. 115–119 (2004)
42. Wenchao, M., Lianchen, L., Anyan, C.: A comparative study on Chinese word segmentation using statistical models. In: Proceedings of IEEE International Conference on Software Engineering and Service Sciences (ICSESS), pp. 482 – 486 (2010)
43. Saha, S.K., Sarkar, S., Mitra, P.: A hybrid feature set based maximum entropy Hindi named entity recognition. In: Proceedings of the IJCNLP 2008 Workshop on NLP for Less Privileged Languages, pp. 343–349 (2008)
44. Ekbal, A., Haque, R., Das, A., Poka, V., Bandyopadhyay, S.: Language independent named entity recognition in Indian languages. In: Proceeding of International Joint Conference on Natural Language Processing (IJCNLP), pp. 1–7 (2008)
45. Cover, T.M., Hart, P.E.: Nearest neighbour pattern classification. IEEE Trans. Inf. Theory **13**(1), 21–27 (1967)
46. Peterson, L.: K-nearest neighbor. Scholarpedia **4**, 1883 (2009)
47. Lifshits, Y.: Nearest neighbor search. In: SIGSPATIAL, Vol. **2**, p. 12 (2010)
48. Agrawal, V., et al.: Application of k-NN regression for predicting coal mill related variables. In: 2016 International Conference on Circuit, Power and Computing Technologies (ICCPCT), India, pp. 1–9 (2016)
49. Qin, Z., Wang, A.T., Zhang, C., Zhang, S.: Cost-sensitive classification with k-nearest neighbors. In: Wang, M. (ed.) KSEM 2013. LNCS (LNAI), vol. 8041, pp. 112–131. Springer, Heidelberg (2013). https://doi.org/10.1007/978-3-642-39787-5_10
50. Hinton, G., Salakhutdinov, R.: Reducing the dimensionality of data with neural networks. Science **313**(5786), 504–507 (2006)
51. Han, J., Kamber, M., Pei, J.: Data Mining: Concepts and Techniques. Elsevier, Amsterdam (2006)
52. Vedala, R. et al.: An application of Naive Bayes classification for credit scoring in e-lending platform. In: ICDSE, pp. 81–84 (2012)

53. Sunny, S., David Peter, S., Jacob, K.P.: Combined feature extraction techniques and Naive Bayes classifier for speech recognition. In: Computer Science & Information Technology (CS & IT), pp. 155–163, (2013)
54. Bahl, L.R., de Souza, P.V., Gopalakrishnan, P.S., Nahamoo, D., Picheny, M.A.: Context dependent modeling of phones in continuous speech using decision trees. In: Proceedings DARPA Speech and Natural Language: Proceedings of a Workshop, pp 264–270, Pacific Grove, Calif (1991)
55. Kotsiantis, S.B.: Decision trees: a recent overview. Artif. Intell. Rev. **39**, 261–283 (2013)
56. Boros, T., Dimitrescu, S.D., Pipa, S.: Fast and accurate decision trees for natural language processing tasks. In: Proceedings of Recent Advances in Natural Language Processing, Sep. 4–6, Varna, Bulgaria, pp. 103–110 (2017)
57. Ekbal, A., Bandyopadhyay, S.: Named entity recognition in Bengali: a multi-engine approach. Proc. Northern Eur. J. Lang. Tech. **1**, 26–58 (2009)
58. Antony, P., Mohan, S.P., Soman, K.: SVM based part of speech tagger for Malayalam. In: Proceedings of IEEE International Conference on Recent Trends in Information, Telecommunication and Computing (ITC), pp. 339–341 (2010)
59. Brin, S.: Extracting patterns and relations from the world wide web. In: Atzeni, P., Mendelzon, A., Mecca, G. (eds.) WebDB 1998. LNCS, vol. 1590, pp. 172–183. Springer, Heidelberg (1999). https://doi.org/10.1007/10704656_11
60. Agichtein, E., Gravano, L.: Snowball: extracting relations from large plain-text collections. In: Proceedings of the Fifth ACM Conference on Digital Libraries, pp. 85–94 (2000)
61. Batista, D.S, Martins, B., Silva, M.J.: Semi-supervised bootstrapping of relationship extractors with distributional semantics. In: Empirical Methods in Natural Language Processing. ACL (2015)
62. Bindal, A., Pathak, A.: A survey on k-means clustering and web-text mining. IJSR **5**(4), 1049–1052 (2016)
63. Sun, J.: Clustering algorithms research. J. Software **19** (2008)
64. Bouhmala, N.: How good is the euclidean distance metric for the clustering problem. In: IIAI-AAI, Kummamoto, pp. 312–315 (2016)
65. Gersho, A., Gray, R.M.: Vector Quantization and Signal Compression, 6th edn. Kluwer Academic Publishers, Boston (1991)
66. Makhoul, J., et al.: Vector quantization in speech coding. In: Proceedings of the IEEE, Vol. 73, no. 11, pp. 1551–1588, November 1985
67. Linde, Y., Buzo, A., Gray, R.M.: An algorithm for vector quantization. In: IEEE COM-28, no. 1, pp. 84–95, January 1980
68. Spanias, A.S.: Speech coding: a tutorial review. In: Proceedings of the IEEE, vol. 82, no. 10, pp. 1441–1582, October 1994
69. Spanias, A., Painter, T., Atti, V.: Audio Signal Processing and Coding. Wiley, New York (2007)
70. Kim, Y.: Convolutional neural networks for sentence classification. In: Conference on Empirical Methods in Natural Language Processing, pp. 1746–1751 (2014)
71. Yin, W., Schiitze, H.: Multichannel variable-size convolution for sentence classification. In: Conference on Computational Language Learning, pp. 204–214 (2015)
72. Burukin, S.: NLP-based data preprocessing method to improve prediction model accuracy (2019). https://towardsdatascience.com/nlp-based-data-preprocessing-method-to-improve-prediction-model-accuracy-30b408a1865f. Accessed 10 May 2020
73. Ding, C., He, X., Zha, H., Simon, H.D.: Adaptive dimension reduction for clustering high dimensional data. In: IEEE International Conference on Data Mining, pp. 147–154 (2002)
74. Guillaumin, M., Verbeek, J.: Multimodal semi-supervised learning for image classification. In: Computer Vision and Pattern Recognition, pp. 902–909 (2010)

75. Kulis, B., Basu, S.: Semi-supervised graph clustering: a kernel approach. Mach. Learn. **74**, 1–22 (2009)
76. Zhou, Z.H., Li, M.: Semi-supervised regression with co-training. In: International Joint Conferences on Artificial Intelligence, pp. 908–913 (2005)
77. Chen, P., Jiao, L.: Semi-supervised double sparse graphs based discriminant analysis for dimensionality reduction. Pattern Recogn. **61**, 361–378 (2017)
78. Bennett, K.P., Demiriz, A.: Semi-supervised support vector machines. In: Advances in Neural Information Processing Systems, pp. 368–374 (1999)
79. Cheung, E.: Optimization Methods for Semi-Supervised Learning. University of Waterloo (2018)
80. Chapelle, O., Sindhwani, V., Keerthi, S.S.: Optimization techniques for semi-supervised support vector machines. J. Mach. Learn. Res. **9**, 203–233 (2008)
81. Li, Y.F., Tsang, I.W.: Convex and scalable weakly labeled SVMs. J. Mach. Learn. Res. **14**, 2151–2188 (2013)
82. Chapelle O., Sindhwani V., Keerthi, S.S.: Branch and bound for semi-supervised support vector machines. In: Advances in Neural Information Processing Systems, pp. 217–224 (2007)
83. Estivill-Castro, V., Yang, J.: Fast and robust general purpose clustering algorithms. In: Mizoguchi, R., Slaney, J. (eds.) PRICAI 2000. LNCS (LNAI), vol. 1886, pp. 208–218. Springer, Heidelberg (2000). https://doi.org/10.1007/3-540-44533-1_24
84. Jolliffe, I.: Principal component analysis. In: International Encyclopedia of Statistical Science, pp. 1094–1096 (2011)
85. Yao, M.: What are important AI & machine learning trends for 2020?. https://www.for bes.com/sites/mariyayao/2020/01/22/what-are–important-ai–machine-learning-trends-for-2020/#3f46f07b2323. Accessed 13 May 2020
86. Zhuang, C., Zhai, A.L., Yamins, D.: Local aggregation for unsupervised learning of visual embeddings. [CS, CV]. https://arxiv.org/abs/1903.12355 (2019)

Improving NER Performance by Applying Text Summarization on Pharmaceutical Articles

Jovana Dobreva, Nasi Jofche, Milos Jovanovik[✉], and Dimitar Trajanov

Faculty of Computer Science and Engineering, Ss. Cyril and Methodius University, Skopje, North Macedonia
jovana.dobreva@students.finki.ukim.mk,
{nasi.jofche,milos.jovanovik,dimitar.trajanov}@finki.ukim.mk

Abstract. Analyzing long text articles in the pharmaceutical domain, for the purpose of knowledge extraction and recognizing entities of interest, is a tedious task. In our previous research efforts, we were able to develop a platform which successfully extracts entities and facts from pharmaceutical texts and populates a knowledge graph with the extracted knowledge. However, one drawback of our approach was the processing time; the analysis of a single text source was not interactive enough, and the batch processing of entire article datasets took too long. In this paper, we propose a modified pipeline where the texts are summarized before the analysis begins. With this, the source articles is reduced significantly, to a compact version which contains only the most commonly encountered entities. We show that by reducing the text size, we get knowledge extraction results comparable to the full text analysis approach and, at the same time, we significantly reduce the processing time, which is essential for getting both real-time results on single text sources, and faster results when analyzing entire batches of collected articles from the domain.

Keywords: Named entity recognition · Data processing · Text summarization · Knowledge extraction · Knowledge graphs

1 Introduction

Large quantities of text from the pharmaceutical domain are available on the Web, and they can be used to build and improve domain-specific natural language processing (NLP) models. These models can then be used for understanding and extracting knowledge specific for the pharmaceutical domain.

These texts can be analyzed using several techniques. Named entity recognition (NER) is a technique used for detecting and recognizing named entities in a text [18]. Multiple pretrained models, which are able to recognize general entities, such as a `Person`, `City`, `Object`, etc., are available on the Web. Our main research interest lies in recognizing entities in the pharmaceutical domain, such as entities which represent a `Pharmaceutical Organization` or a `Drug`. In our

© Springer Nature Switzerland AG 2020
V. Dimitrova and I. Dimitrovski (Eds.): ICT Innovations 2020, CCIS 1316, pp. 87–97, 2020.
https://doi.org/10.1007/978-3-030-62098-1_8

previous research efforts [9] we developed a platform which extracts entities and facts from pharmaceutical texts, and then populates a knowledge graph with the extracted knowledge[1]. This approach had one drawback: the processing time. In it, the analysis of a single text source was not interactive enough, and the batch processing of entire article datasets took too long.

Today, all state-of-the-art models for NLP are based on machine learning (ML). These models are mainly based on the use of bidirectional long-short term memory (LSTM) networks [10], convolutional networks [24] and lately, the use of transformer architectures for improving the accuracy of NER. If we analyze the grammatical aspect of the English language, it is evident that it is a set of rules, just like all other languages. On the other hand, the semantic aspect of the sentence is non-deterministic and depends on the author's style of writing. In other words, NER uses the grammatical structure to highlight all entities in a given sentence. The entities are followed by a post tag, which is a brief description of the grammatical meaning of the word. The key emphasis is the automated implementation of common language processing tasks, such as tokenization, dealing with punctuation and stop terms, lemmatization, as well as the possibility of custom, business case specific text processing functions, such as joining consecutive tokens to tag a multi-token object, or perform text similarity computations. Two well-known language processing libraries, spaCy [8] and AllenNLP [7], which come with a pretrained model based on convolutional layers with residual connections and a pretrained model based on Elmo embedding, respectively, evaluate the overall applicability and accuracy of this technique.

The occurrence of pretrained models stems from the need for good accuracy by highlighting the entities. Therefore, BERT [6] has emerged as one of the most well-known language representation models. Unlike other models, BERT is designed to pretrain profound bidirectional representations from unlabeled text, by jointly conditioning in all layers on both the left and right sense [21]. As a result, the pretrained BERT model can be fine-tuned to create state-of-the-art models for a wide variety of activities, such as answering questions and language inference, without substantial task-specific architecture modifications.

As we mentioned above, NER highlights grammatically correct entities. Therefore, the underlying binding phenomena in the field of syntax is established with co-reference resolution. Semantic role of words and phrases in the sentence is supervised by semantic role labeling (SRL). This is an important step towards making sense of the meaning of a sentence. In other words, SRL represents the process that assigns labels to words or phrases in a sentence, which indicate their semantic role, such as that of an agent, a goal, or a result.

The final purpose is, with the help of knowledge extraction, to generate a knowledge graph (KG), using the Resource Description Framework (RDF) - a graph-oriented knowledge representation of the entities and their semantic relations. The generated RDF graph dataset represents a collection of multiple knowledge extractions, processed from different text sources, and at the same time links the extracted entities to their counterparts within DBpedia and the

---

[1] https://github.com/f-data/NER_Pharma.

rest of the Linked Data on the Web [1]. This provides access over the entire extracted knowledge and the relevant linked knowledge already present in the publicly available knowledge graphs.

In this paper we look at the possibility of applying summarization on each source text, i.e. article from the pharmaceutical domain, to get a simplified and shortened version which still contains all important entities from it. Our results show that all of the above-mentioned techniques work equally or better on summarized texts than when used on full texts, depending on how frequently the objects of interest appear in the sentences. With this, we get more effective NER processing and, at the same time, we significantly reduce the processing time, which is essential for getting both real-time results on single text sources, and faster results when analyzing entire batches of collected texts from the domain.

2   Related Work

The development of technology continues to grow exponentially, which is especially visible in the fields of NLP and ML which are becoming more and more ubiquitous. This is the reason for the large amount of research being done and research papers being published in recent time, which impacts the top ranking of the state-of-the-art models and makes the list rather dynamic.

One popular approach is the process of label sequencing with neural networks, which outperform early NER systems based on domain dictionaries, lexicons, orthographic feature extraction and semantic rules [4]. This approach is then outperformed by convolutional neural networks and the representation of words with a n-dimensional matrix [24]. Models based on LSTM networks [20] are also popular today, and one of the reasons for why this model is so well-known, is the fact that it is a part of the BERT model. As we already noted, BERT is one of the latest milestones in the development of this field [6]. It is based on a transformer architecture and integrates an attention mechanism. The transformers are built from a LSTM-encoder and a LSTM-decoder. The main idea behind the attention technique is that it allows the decoder to *look back* at the complete input and extracts significant information that is useful in decoding [3].

There are different ways to construct an RDF-based knowledge graph, which generally depend on the source data [16]. In our case, we work with extracted and labeled data, so we can utilize existing solutions which recognize and match the entities in our data with their corresponding version in other publicly available knowledge graphs.

There are many platforms, such as AllenNLP [7] and spaCy [8], which aim to provide demo pages for NLP model testing, along with code snippets for easier usage by machine learning experts. On the other hand, projects like Hugging Face' Transformers [23] and Deep Pavlov AI [2] are libraries that significantly speed up prototyping and simplify the creation of new solutions based on the existing NLP models.

In this paper, we focus on simplicity and we use the gensim library[2] [17] for the implementation of text summarization. This open library uses the words in the text to create a graph comprised of them, where they apply TextRank [15] for making a weighted graph. Intuitively, TextRank works well because it does not only rely on the local context of a text unit (vertex), but rather it takes into account information recursively drawn from the entire text (graph). The text units selected by TextRank for a given application are the ones most recommended by related text units in the given article, with preference given to the recommendations made by the most influential ones, i.e. the ones that are in turn highly recommended by other related units. Nevertheless, according to the knowledge we have, there is still a gap in finding common solutions for knowledge extraction in the pharmaceutical domain that is human-centric and enables visualization of the results in a human-understandable format. In this paper, besides knowledge extraction, we seek to optimize the problem by summarizing long news articles from the pharmaceutical domain.

3   Text Analysis and Knowledge Extraction

Initially, we classify whether a given text originates from the pharmaceutical domain, and only the ones which are positively graded are taken into account for further analysis. Every properly classified pharmaceutical text is further analyzed by identifying combined entities via the proposed models, as well as by using BioBERT to detect BC5CDR4 and BioNLP13CG5 tags [22] including disease, chemical, cell, organ, organism, gene, etc. In addition, we use a fine-tuned BioBERT model to identify pharmaceutical organizations and drugs, groups of individuals that are not covered by the standard NER tasks.

Recognized entities act as a benchmark for identifying their references in the entire text by applying context co-reference resolution: the replacement of references ("it", "it's", "his", etc.) with the individual they refer to. Libraries such as AllenNLP, StanfordNLP [12] and NeuralCoref[3] include co-reference resolution algorithms. After resolving this linguistic problem, we apply semantic role labeling on the text.

As a final step, with the help of state-of-the-art models for knowledge extraction and DBpedia Spotlight [13], we annotate the text and RDF statements, representing the knowledge in the text, are generated. The results obtained are then enriched with additional RDF facts which we construct from the Pharmaceutical Organization and Drug entities listed. This enriched RDF knowledge graph is then available inside or outside of the system for further use and analysis.

4   Analysis and Results

The general idea of this research paper comes down to being able to extract the same amount of context, or at least the essential part, from a summarized version of a long text article. By initially being based on our PharmKE data platform, which is able to create RDF knowledge graphs from recognized pharmaceutical entities in long texts[4], we go one step further by improving performance of the entire text analysis pipeline by showing that similar, and sometimes better results can be obtained by analyzing a previously summarized text. To achieve this, we use the gensim library [17], which is based on the TextRank algorithm [14] - a graph based ranking algorithm for graphs extracted from natural language texts. The generated graph from a text interconnects words or other text entities with meaningful relations. Text units of various sizes and characteristics may be inserted as vertices in the graph, e.g. terms, collocations, whole sentences, or others, depending on the application in hand. Similarly, the specification determines the form of relationships used to draw links between the two vertices of this kind, e.g.. lexical or semantic relationships, contextual overlap, etc. The application of graph-based ranking algorithms to natural language texts [18] consists of the following main steps:

1. Identifying the units of text that best represent the task at hand, and attaching them to the graph as vertices.
2. Identifying the relationships that link these units of text, and using those relationships to draw edges in the graph between vertices. Edges can be directed or undirected, weighted or unweighted.
3. Iterate the graph-based ranking algorithm until convergence.

Other sentence similarity measures, such as string kernels, cosine similarity, longest common sub-sequence, etc. are also possible, and we are currently evaluating their impact on the summarization. The text is therefore represented as a weighted graph, and consequently is using the weighted graph-based ranking formula. After the ranking algorithm is run on the graph, sentences are sorted in reversed order of their score, and the top ranked sentences are selected for inclusion in the summary.

Here, we show an example output of the summarization algorithm, when applied on a given pharmaceutical text. As we can see, the summarized version of the text contains most of the context of the original pharmaceutical article:

– Original text: *"It may not seem as if pharma companies are biting their nails over drug prices. After all, a cohort of drugmakers made headlines last week with an annual round of January hikes. But the issue is on the top of industry worries for 2019, a new report finds. Slightly more than half of respondents to a GlobalData survey tagged pricing and reimbursement as their biggest worry this year, the analytics firm said. As that pricing pressure rolls on, GlobalData analysts also expect aggressive negotiation tactics to drive down drug prices."*.
– Summarized version of the same text: *"Slightly more than half of respondents to a GlobalData survey tagged pricing and reimbursement as their biggest worry this year, the analytics firm said."*.

---

[4] The articles were retrieved from https://www.fiercepharma.com/, https://www.pharmacist.com/ and https://www.pharmaceutical-journal.com/.

Table 1. Comparison of the average named entity recognition results between the original and summarized articles using TextRank summarization.

		Original		Summarized		Coverage		
	Size range	Spotlight entities (Avg)	Additional entities (Avg)	Spotlight entities (Avg)	Additional entities (Avg)	Spotlight entities	Additional entities	Total entities
Group 1	[1, 4) KB	8.80	4.66	4.94	3.25	71%	95%	72%
Group 2	[4, 6) KB	9.03	6.75	5.00	3.11	68%	74%	65%
Group 3	[6, 18] KB	10.00	12.97	5.38	4.23	62%	62%	55%
Total	[1, 18] KB	8.99	6.16	5.00	3.27	69%	84%	68%

This methodology is applied to 1,500 pharmaceutical articles that are part of the collected dataset, and then, by utilizing the PharmKE platform, we extract knowledge and build RDF triples. The PharmKE platform is able to expand the set of named entities initially detected by DBpedia Spotlight [5] by a substantial percentage, thus improving the overall entity extraction. This is done by additionally recognizing Drug and Pharmacy Organization entities. The results in Table 1 and Table 2 show that almost all of the entities recognized in the original texts, are recognized in the summarized articles, as well. This is shown via the coverage measure: it denotes the percentage of entities recognized in summarized articles, compared to the ones recognized in the original ones.

Table 1 shows our results when we used TextRank summarization, while Table 2 shows our results using latent semantic analysis (LSA) summarization [19]. As we can see, LSA summarization provides a better coverage of the original entities.

Table 2. Comparison of the average named entity recognition results between the original and summarized articles using LSA summarization.

		Original		Summarized		Coverage		
	Size range	Spotlight entities (Avg)	Additional entities (Avg)	Spotlight entities (Avg)	Additional entities (Avg)	Spotlight entities	Additional entities	Total entities
Group 1	[1, 4) KB	8.80	4.66	6.72	4.47	93%	121%	94%
Group 2	[4, 6) KB	9.03	6.75	6.53	4.10	84%	93%	80%
Group 3	[6, 18] KB	10.00	12.97	6.44	3.68	71%	48%	55%
Total	[1, 18] KB	8.99	6.16	6.61	4.26	88%	104%	85%

These results show that by applying the summarization technique, we are still able to keep a large amount of core entities in the shorter text, while reducing the text processing time at the same time - thus significantly improving the performance of the analysis, as shown in Table 3 and 4.

Table 3. Comparison of the average processing time between the original and summarized articles using TextRank summarization.

	Original			Summarized		Speed-up
	Size range	Size (Avg)	Processing time (Avg)	Size (Avg)	Processing time (Avg)	
Group 1	[1, 4) KB	2.82 KB	**19.53 s**	1.33 KB	**8.25 s**	**2.36**
Group 2	[4, 6) KB	4.23 KB	**19.43 s**	1.39 KB	**12.93 s**	**1.50**
Group 3	[6, 18] KB	7.42 KB	**18.34 s**	1.28 KB	**12.04 s**	**1.52**
Total	[1, 18] KB	3.75 KB	**19.39 s**	1.35 KB	**12.20 s**	**1.59**

As we can see from Table 3 and Table 4, the speed-up varies from 1.48x–2.67x between different groups of articles (grouped by size), and the average speed-up is 1.59 and 1.58. This is a significant performance improvement. This improvement is relevant for both single text analytics, as it provides more interactive response times, as well as for batch analysis of a large text corpus.

Table 4. Comparison of the average processing time between the original and summarized articles using LSA summarization.

	Original			Summarized		Speed-up
	Size range	Size (Avg)	Processing time (Avg)	Size (Avg)	Processing time (Avg)	
Group 1	[1, 4) KB	2.82 KB	**19.53 s**	1.09 KB	**11.65 s**	**2.67**
Group 2	[4, 6) KB	4.23 KB	**19.43 s**	1.07 KB	**13.06 s**	**1.48**
Group 3	[6, 18] KB	7.42 KB	**18.34 s**	1.11 KB	**12.13 s**	**1.51**
Total	[1, 18] KB	3.75 KB	**19.39 s**	1.09 KB	**12.25 s**	**1.58**

The results in Table 1 and Table 2 show that in some cases simplicity is better for named entity recognition. However, we should note that these numbers also depend on the number of words. Namely, if an article has more entities, as is the case with the original full texts, it is more likely that the error level will increase. This means that gensim allows us to increase the accuracy of labeling, because the summarized texts include only the important entities from the pharmaceutical domain. On the other hand, the latent semantic analysis is appropriate for text summarization, because LSA uses a term-document matrix which describes the occurrences of terms in a particular document. It is a sparse matrix whose rows correspond to terms and whose columns correspond to documents/topics.

From the documentation of the TextRank algorithm, it is obvious that the word is represented by its own index. Therefore, the number of the edges between words is proportional with the number of different words. In other words, the weights of the edges represent the probability of the pair of words between the given edge.

On the other side, LSA is making an occurrence matrix, where the weight of an element of the matrix is proportional to the number of times the terms appear in each topic, where rare terms are upweighted to reflect their relative importance.

Meanwhile, the summarized sentences are likely to contain many of the entities that are part of our interest, as the summarization is performed with regards to the most important words from the context. On the other hand, all phrases that are inconsequential will be removed, and both of these actions lead to an increase in the accuracy of highlighting the right entities and building exact RDF triples.

With these results, we show that significant improvements in performance are established, while keeping the consistency of the generated knowledge graph, and in some cases even expanding it. By making the articles significantly shorter, we are able to speed up the text processing pipeline in our system (Fig. 1), with no effect on the output quality.

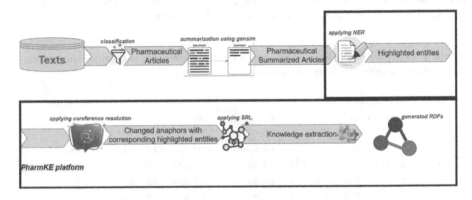

Fig. 1. Diagram of our processing approach

5   Conclusion

The main goal of this research is to show how shortening the input text can have a tremendous impact on the performance of the analytical pipeline, while at the same time preserving the quality of the knowledge extraction results. Through our pipeline comprised of NER, co-reference resolution and SRL, we were able to prove this hypothesis over a dataset of 1,500 articles from the pharmaceutical domain. We showed that we can significantly shorten the processing time and speed up the process, all while keeping the same level of quality of the knowledge extraction. This improvement is relevant for both single text analytics, as it provides more interactive response times, as well as for batch analysis of a large text corpus.

The results show that the NER improvements are persistent when summarizing the source articles, albeit the absolute number of detected entities is smaller. This is an expected trade-off, which we wanted to point out.

The need to simplify pharmaceutical texts came from the idea to generate grammatically simplified sentences as a way of extracting simple questions and providing concise answers. In other words, the summarization extracts the most frequent parts of a given text by generating all the relevant questions and providing answers that consist of the most substantial information about the entities. Therefore, when we put the generated questions in the question-answering system [11] we get the correct answers, resulting in a simplified text, comprehensive questions, and short and concise answers. The most far-reaching result we want to achieve is to obtain new attributes that will describe the entities in the pharmaceutical domain through the generated questions and the relevant answers required, allowing us to delineate the knowledge graph and thereby obtain better results for further research. In order to achieve better results in the future, we plan to try different techniques for summarizing the articles, which would further be generalized and applicable outside the pharmaceutical field, as well.

This work is part of our PharmKE platform, a platform for knowledge extraction in the pharmaceutical domain, which uses visualization and helps end-users build knowledge graphs from text sources of their interest. Further enhancements are planned, in order to contribute to further participation in this domain, which in the bright future holds hope for faster development of the pharmaceutical industry, side by side with the IT industry.

Acknowledgement. The work in this paper was partially financed by the Faculty of Computer Science and Engineering, Ss. Cyril and Methodius University in Skopje.

References

1. Bizer, C., Heath, T., Idehen, K., Berners-Lee, T.: Linked data on the web. In: Proceedings of the 17th International Conference on World Wide Web, WWW 2008, pp. 1265–1266. ACM, New York (2008). https://doi.org/10.1145/1367497. 1367760, http://doi.acm.org/10.1145/1367497.1367760

2. Burtsev, M., et al.: DeepPavlov: open-source library for dialogue systems. In: Proceedings of ACL 2018, System Demonstrations, Melbourne, Australia, pp. 122–127. Association for Computational Linguistics, July 2018. https://doi.org/10.18653/v1/P18-4021, https://www.aclweb.org/anthology/P18-4021

3. Chiu, J.P., Nichols, E.: Named entity recognition with bidirectional LSTM-CNNs. Trans. Assoc. Comput. Linguist.**4**, 357–370 (2016). https://doi.org/10. 1162/tacl_a_00104,https://www.aclweb.org/anthology/Q16-1026

4. Collobert, R., Weston, J., Bottou, L., Karlen, M., Kavukcuoglu, K., Kuksa, P.: Natural language processing (almost) from scratch. J. Mach. Learn. Res. **12**, 2493–2537 (2011)

5. Daiber, J., Jakob, M., Hokamp, C., Mendes, P.N.: Improving efficiency and accuracy in multilingual entity extraction. In: Proceedings of the 9th International Conference on Semantic Systems, pp. 121–124. Association for Computing Machinery (2013)

6. Devlin, J., Chang, M.W., Lee, K., Toutanova, K.: BERT: pre-training of deep bidirectional transformers for language understanding. arXiv preprint arXiv:1810.04805 (2018)

7. Gardner, M., et al.: AllenNLP: A deep semantic natural language processing platform. In: Proceedings of Workshop for NLP Open Source Software (NLP-OSS), Melbourne, Australia, pp. 1–6. Association for Computational Linguistics, July 2018. https://doi.org/10.18653/v1/W18-2501, https://www.aclweb.org/anthology/W18-2501

8. Honnibal, M., Montani, I.: spaCy 2: Natural Language Understanding with Bloom Embeddings. Convolutional Neural Networks and Incremental Parsing (2017, to appear)

9. Jofche, N.: Master's thesis: analysis of textual data in the pharmaceutical domain using deep learning. Faculty of Computer Science and Engineering (2019)

10. Kuru, O., Can, O.A., Yuret, D.: CharNER: character-level named entity recognition. In: Proceedings of COLING 2016, the 26th International Conference on Computational Linguistics: Technical Papers, The COLING 2016 Organizing Committee, Osaka, Japan, pp. 911–921. December 2016. https://www.aclweb.org/anthology/C16-1087

11. Lamurias, A., Couto, F.M.: LasigeBioTM at MEDIQA 2019: biomedical question answering using bidirectional transformers and named entity recognition. In: Proceedings of the 18th BioNLP Workshop and Shared Task, Florence, Italy, pp. 523–527. Association for Computational Linguistics, August 2019. https://doi.org/10.18653/v1/W19-5057, https://www.aclweb.org/anthology/W19-5057

12. Manning, C., Surdeanu, M., Bauer, J., Finkel, J., Bethard, S., McClosky, D.: The stanford CoreNLP natural language processing toolkit. In: Proceedings of 52nd Annual Meeting of the Association for Computational Linguistics: System Demonstrations, Baltimore, Maryland, pp. 55–60. Association for Computational Linguistics, June 2014. https://doi.org/10.3115/v1/P14-5010, https://www.aclweb.org/anthology/P14-5010

13. Mendes, P.N., Jakob, M., Garcia-Silva, A., Bizer, C.: DBpedia spotlight: shedding light on the web of documents. In: Proceedings of the 7th International Conference on Semantic Systems (I-Semantics). Association for Computing Machinery (2011)

14. Mihalcea, R., Tarau, P.: TextRank: bringing order into texts. In: Proceedings of EMNLP-04 and the 2004 Conference on Empirical Methods in Natural Language Processing. Association for Computational Linguistics, July 2004

15. Mihalcea, R., Tarau, P.: TextRank: bringing order into text. In: Proceedings of the 2004 Conference on Empirical Methods in Natural Language Processing, pp. 404–411. Association for Computational Linguistics (2004)

16. Paulheim, H.: Knowledge graph refinement: a survey of approaches and evaluation methods. Semantic Web 8(3), 489–508 (2017)

17. Řehůřek, R., Sojka, P.: Software framework for topic modelling with large Corpora. In: Proceedings of the LREC 2010 Workshop on New Challenges for NLP Frameworks, ELRA, Valletta, Malta, pp. 45–50, May 2010. http://is.muni.cz/publication/884893/en

18. Srinivasa-Desikan, B.: Natural Language Processing and Computational Linguistics: A Practical Guide to Text Analysis with Python, Gensim, SpaCy, and Keras. Expert insight, Packt Publishing (2018). https://books.google.mk/books?id=_tGctQEACAAJ

19. Steinberger, J., Ježek, K.: Using latent semantic analysis in text summarization and summary evaluation. In: Proceedings of the ISIM 2004, pp. 93–100 (2004)

20. Sundermeyer, M., Schlüter, R., Ney, H.: LSTM neural networks for language modeling. In: Thirteenth Annual Conference of the International Speech Communication Association (2012)

21. Vaswani, A., et al.: Attention is all you need. In: Guyon, I., et al. (eds.) Advances in Neural Information Processing Systems, vol. 30, pp. 5998–6008. Curran Associates, Inc. (2017). http://papers.nips.cc/paper/7181-attention-is-all-you-need.pdf

22. Wang, X., et al.: Cross-type biomedical named entity recognition with deep multi-task learning. Bioinformatics **35**(10), 1745–1752 (2019). https://doi.org/10.1093/bioinformatics/bty869

23. Wolf, T., et al..: Hugging face's transformers: state-of-the-art natural language processing. ArXiv abs/1910.03771 (2019)

24. Zhu, F., Shen, B.: Combined SVM-CRFs for biological named entity recognition with maximal bidirectional squeezing. PLoS One **7**, 39230 (2012)

Case Study: Predicting Students Objectivity in Self-evaluation Responses Using Bert Single-Label and Multi-Label Fine-Tuned Deep-Learning Models

Vlatko Nikolovski[✉], Dimitar Kitanovski, Dimitar Trajanov,
and Ivan Chorbev

Faculty of Computer Science and Engineering, Ss. Cyril and Methodius University in
Skopje, Rugjer Boshkovikj 16, 1000 Skopje, Republic of North Macedonia
{vlatko.nikolovski,dimitar.trajanov,ivan.chorbev}@finki.ukim.mk,
dimitar.kitanovski@students.finki.ukim.mk

Abstract. Students' feedback data regarding teachers, courses, teaching tools, and methods represent valuable information for the education system. The obtained data can contribute in enhancing and improving the education system. Feedback from students is of great essence in the process of extracting hidden knowledge using various techniques for data mining and knowledge discovery. This paper presents various tools and methods for analyzing students' feedback using Sentiment and Semantic analyses. The essential task in Sentiment analysis is to extract the particular sentiment from textual student responses in terms of negative and positive reactions, while the Semantic analysis contributes to combine textual response in the specific group based on questions. The output produced by the Sentiment and Semantic analyses provides a direct relationship between qualitative and quantitative parts of the evaluation in the form of student comments and grades.

Keywords: Natural language processing · Machine learning · Deep learning · Text mining · Bert · Data processing · Data analyses

1 Introduction

Self-evaluation systems are increasingly becoming a trend in higher education. The self-evaluation systems collect a lot of data that can be used for Data Mining, resulting in a method also known as Educational Data Mining [11]. Furthermore, educational data is of great importance and is widely used in Learning Analytics [5] area as well.

Data Mining is a broad research area supported by a number of tools and techniques used to turn raw data into useful information. One of the domains in Data Mining is Natural Language Processing (NLP) [4]. NLP includes various methods that enable computer systems to understand and manipulate natural

© Springer Nature Switzerland AG 2020
V. Dimitrova and I. Dimitrovski (Eds.): ICT Innovations 2020, CCIS 1316, pp. 98–110, 2020.
https://doi.org/10.1007/978-3-030-62098-1_9

language text or speech to do compelling analyses. Today NLP is becoming increasingly available and exact thanks to the huge improvements in data access as well as the increase in computational power that provides meaningful results in multitude areas, such as Medicine, Finance and Education, and many more. There are many natural language processing models developed so far, but this study focuses mostly on Bert [6] that is one of the newest and most accurate models for NLP tasks.

Bert (Bidirectional Encoder Representations from Transformers) [6] represents the NLP pre-trained model developed by Google using generic data-sets (from Wikipedia and BooksCorpus) that can be used to solve different NLP tasks. The pre-trained Bert model can be fine-tuned with additional output layers allowing scientists to create state-of-the-art models for many various NLP tasks, such as Single-Label and Multi-Label classification. Besides, Sentiment analysis is a form of a Single-Label fine-tuning task on Bert that classifies a text to understanding the emotions expressed. While Multi-Label fine-tuning task on Bert provides a mechanism for text classification over multiple predefined class categories.

This case study presents an educational data mining technique for prediction of students' objectivity towards grading courses, teachers, teaching tools and methods. In practice, students providing objective grading is of great importance to universities, because the improvement of academic tools and methods is strictly tied to the feedback from the students and their academic success. In order to evaluate the objectivity, this case study presents a model that examines the free-text comments from the data-set using natural language understanding. Also, the comments are analyzed with fine-tuned Bert Sentiment classification models to understand the sentiment expressed. The comments are then analyzed with a fine-tuned Multi-Label model to classify the comments in predefined groups. Finally, an average grade is calculated for each of the groups provided by the students grading. This gives an overview of the relationship between the comments and grades provided by the students that pose as estimator of students' feedback.

2   Related Work

Education is one of the multitudes of areas that offer different and rich data-sets as well as opportunities to extract knowledge from those data-sets. Such educational data is collected through educational software that is the main link between students and the educational institution. There are many such educational software applications, developed in order to facilitate the whole educational process, as well as to digitize the large amount of information necessary for flawless work. Some of these education software applications are presented in [1] and [7] where authors describe various modules offered by those software.

Data mining techniques have gained momentum in the last decade. Different applications, tools, and future insights from the educational data mining field are presented in [12]. On the other hand, Deep Learning and Learning Analytics have become key techniques used to extract knowledge from raw data, especially in the education field. Authors in [8] have presented comparison between different estimation models applied on the education data. All of these models are developed to explore the utility and applicability of deep learning for educational data mining and learning analytics. Also in [13] and [3] are presented various techniques that are used to predict academic student performance. In [13] a deep neural network is deployed to anticipate the risk category of students in order to provide measures for early intervention in such cases, while in [3] a decision tree based model is provided whose main task is classification. Furthermore, such a developed model is used to evaluate students performance. Besides, authors in [10] initially have presented an overview of several data mining algorithms, whose main purpose is analysis of students data and dropout prediction. Moreover, authors provide a predictive model which will identify a subset of students who tend to drop out of the studies after the first year.

Natural language processing is a sub-field in the area of Deep Learning and offers a variety of tools that are used to process natural languages/text. One of the most famous pre-trained model/tool used in the field of natural language processing is the model developed from Google, known as Bert (Bidirectional Transformers for Language Understanding) [6]. The usage of natural language processing within education is described in [9]. Besides, authors in [2] describe tools for understanding text from free-style answers in course evaluation forms. Furthermore, authors introduced "Teaching evaluation index (TEI)" to evaluate textual evaluations from students using the number of positive and negative comments obtained from the free-text questions.

3   Data-Set Description

This study examines a students' feedback dataset obtained from the Ss Cyril and Methodius University in Skopje. All students responses are collected using the self-evaluation system developed in house by staff from the Faculty of Computer Science and Engineering in Skopje, respectively.

The survey contains an identical set of 12 questions for every course, on which students can leave their responses. The collection of questions is characterized by 11 quantitative questions with answers valued on a scale of 5 to 10 (hereafter referred to as "GRADES") and one qualitative question in the form of a free text comment (hereafter referred as "FREETEXT") where students can leave their remarks, critiques, praise or opinions regarding the specific course. The focus of this study is to establish a model that can explain the relationship between GRADES and FREETEXT in the manner of predicting the objectivity of the grading.

The data is organized in a non-relational database, where each record contains information about a specific semester, course, and student answers. The data-set contains students responses for 4 consecutive semesters. Table 4 presents the questions used in the survey, while Table 5 presents a single student feedback for a specific course. Table 1 shows the statistical characteristics or the data-set organized by semester.

Table 1. Data-set description

	Summer semester (2017/2018)		Winter semester (2018/2019)		Summer semester (2018/2019)		Winter semester (2019/2020)	
	GRADE	FREETEXT	GRADE	FREETEXT	GRADE	FREETEXT	GRADE	FREETEXT
Number of records	102,085	18,015	100,405	11,156	118,007	20,824	90,295	22,573

4 Methodology

Various methods and techniques have been conducted for processing students responses to establish further analyses. Natural language processing, context recognition, sentiment, and semantic analysis were just a part of the methods that were used to create the estimation model.

The data-set evaluated in this study contains 2 main subsets:

– GRADE - numerical answers graded on a scale from 5–10
– FREETEXT - free-text answers for expressing students insights

Examining closely a single record R that represents a single student feedback S for a specific course C, it contains 11 grades (on a scale of 5–10) titled GRADE and 1 free text comment FREETEXT. The GRADE part in a context of single R, evaluates several aspects in domain of: teacher, student, teaching tools/methods and student performance. Based on this behaviour, the FREETEXT part should also follow this pattern and give more detailed information/insight in the same domain as GRADE. In order to extract any further significant information from single R, a relationship should be established between the FREETEXT and GRADE parts. The main objective of this relationship is to provide a measurable value SOI (student objectivity index) from FREETEXT part, that can be compared to the average grade of the GRADE part, where the value of the average grade will pose as an estimator of SOI. In addition, single record R expresses high objectivity only when SOI and average grade have a matching value threshold.

To improve the spanning context of the data-set and improve the natural language processing tasks, all 11 questions from the GRADE part are categorized into 4 groups based on the semantic context. Each group contains questions related to a specific context in domain of: teachers, students, teaching tools/methods and students performance. Table 2 shows the distribution of the 11 questions over the 4 different semantic groups.

Table 2. Groups for classification

Group	Questions
Teachers	1, 2, 3, 4
Teaching_tools	6, 7, 8
Examination	9, 10
Students	5, 11

To extract any further information from the free-text comments in domain of the GRADE part, the following objectives should be met:

- Evaluate the semantic context of free-text comments in order to categorize them in a specific semantic group
- Evaluate sentiment from free-text comments in domain of positive/negative values
- Calculate SOI based on the previous analyses

To fulfill the objectives of the relationship, 3 different models are presented for each of the tasks respectively. A Multi-Label classification model was developed that predicts the category of the free-text responses based on the semantic context. The classification layer on top of the Bert network was created over the categories/labels defined in Table 2. This model is capable of predicting the semantic group for every single free-text comment in the data-set. Sentiment classification model was established over students responses to understand the emotion expressed in the manner of negative and positive sentiment. The Average grade model was provided by calculating the average grade from the GRADE part for each semantic group respectively. Finally, with combining all the tasks: sentiment and multi-label classification on the FREETEXT and average grade from the GRADE part, a new model is presented that predicts the student objectivity index (a value that measures the amount of deviation on the grading part). Figure 1 shows the whole process of establishing a model that predicts the student objectivity index.

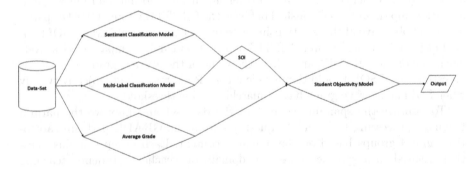

Fig. 1. Process of building a model for student objectivity index

This paper presents three functional models. Sentiment classification and Multi-Label models are derived from Bert [6] with fine-tuning based on the data-set from students' feedback. The first model offers a Single-Label classification layer that can predict the sentiment in free text comments, while the second model contains a Multi-Label classification layer that classifies the free text comments into a specific group presented in Table 2. The third model provides the average grade per group, based on the grades given by the students. Fusion of the Sentiment and Multi-label models provides the Student Objectivity Index value that can be compared to the average grades from the third model, that further evaluates the objectivity of students feedback.

4.1 Data Preparation

The free-text comments from the data-set were translated from Macedonian to English language, in order to train the Sentiment and Multi-label models. Furthermore, the process of filtering was applied over the translated text to exclude irrelevant (very short and not well translated) comments from the data-set. In order to summarise the grades, an average grade was calculated for each of the groups from Table 2 respectively (Fig. 2).

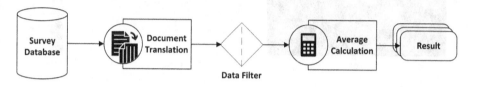

Fig. 2. Data preparation process

4.2 Fine-Tuning Bert - Sentiment Classification

In order to establish an appropriate training model for sentiment analysis, we made a connection between the GRADE and FREETEXT parts of the data-set. A general sentiment classification task will not give correct results unless the network is trained for the specific domain in education. Therefore, the model was custom trained over the specific domain in order to be able to provide the very best understanding of the sentiment context presented in free-text comments. Each student response per course contains 12 answers (11 grades and 1 free text comment). The average grade of the quantitative part is normalized on a scale of 1–5 and then added as a sentiment value to the comment, respectively. Then, the average grade continuous values were replaced with discrete values in terms of positive and negative labels. Data for semesters S2017/18, W2018/18 and S2018/19 was used for training and validation of the Sentiment Classification model, while data for semester W2019/20 was used for prediction.

- Records with an average grade less or equal to 3 corresponds to negative sentiment
- Records with an average grade greater than 3 corresponds to positive sentiment

The trained model for Sentiment analysis has an accuracy of 0.9645. Figure 3 presents the confusion matrix of the model.

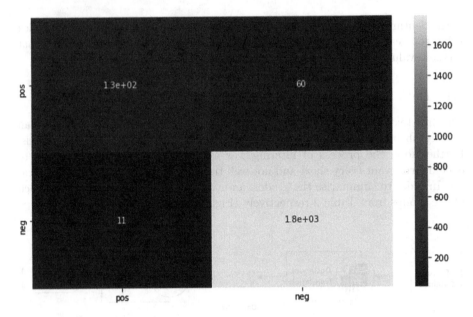

Fig. 3. Confusion matrix of the model for sentiment analysis

4.3   Fine-Tuning Bert - Multi-Label Text Classification

Processing the data for the Multi-Label classification model requires manual labeling of the training set. In addition, the free text comments in the training set were manually divided in one of the four groups presented in Table 2, based on the semantic relationship between the groups and comments. After the manual division of the comments into groups, the train data-set contains discrete values [0, 1] for each of the groups respectively, 1 if the comment belongs to a certain group and 0 otherwise. Data for semesters S2017/18, W2018/18 and S2018/19 was used for training and validation of the Sentiment Classification model, while data for semester W2019/20 was used for prediction.

The trained model for Multi-Label classification has an accuracy of 0.919. Figure 4 presents the confusion matrix of the model.

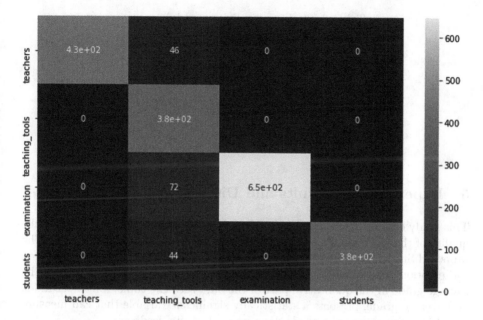

Fig. 4. Confusion matrix of the model for Multi-Label classification

4.4   Student Objectivity Model

Combining the results of Sentiment, Multi-Label Classification, and Average Grade models present the final Student Objectivity model. In addition, fusing the data processed by Sentiment and Multi-Label models define the distribution of sentiments (positive, negative) over the groups defined in Table 2. In order to provide significant statistical value from the distribution, the data is normalized by Formula 1.

$$SOI(norm) = \frac{positive - negative}{positive + negative} \tag{1}$$

Normalization of the sentiments per group provides a statistical value of the distribution on a scale of $[-1, 1]$, where the border values $(-1,1)$ indicate that number of positive sentiments is equal to zero and vice versa.

Finally, the Average Grade model is fused with the normalized sentiment data (SOI). This fusion provides a new dimension of the Student Objectivity model that specifies the relationship between sentiments from free-text comments and grading from FREETEXT and FRADE parts of the data-set respectively. Table 3 presents the fused data-set defined by the Student Objectivity model.

Table 3. Student objectivity model data-set

	Course	Group	SOI	Average Grade
1	25636	Teachers	−1.0	8.973118
2	25336	Teaching-tools	−1.0	9.262136
3	25756	Teachers	1.0	7.824468
4	25637	Examination	1.0	8.824324
5	25339	Students	1.0	8.836247

5 Experimental Results and Discussion

This study estimates the objectivity of students' feedback, an assessment of the quality of the teachers and teaching tools and methods in higher education. The proposed Student Objectivity Model provides a statistical relationship between the emotions expressed in the qualitative (free text comment) part and the average grade in the qualitative part of the data-set. The tuple, sentiment index and average grade, presents a statistically significant variable that can measure the objectivity among the whole data-set of students' feedback.

Sentiment index is a continuous variable in the interval of [−1,1], where −1 indicates a group of data with only negative sentiments while 1 only positive sentiments. Average grade is a continuous variable defined over the interval of [5,10]. Figure 5 shows plotted values of the variables.

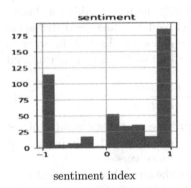

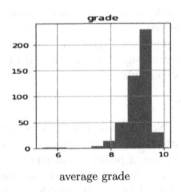

Fig. 5. Histogram: Sentiment index and average grade

The statistical evaluation of the model presents huge discontinuity in the distribution of the tuple *sentiment index, average grade*. Sentiment index variable is transferring from the continuous into the discrete domain set. The evaluation shows that 82% or the values for the sentiment index are converging to the extremes of the interval [−1, 1]. The step of freedom distributed as discrete values of −1 and 1, reveals a pattern: majority of the students were expressing only positive or only negative sentiments among the groups defined in Table 2.

Examination of the average grade shows that 88% of the values are above 8.5. The model reveals another data pattern: students providing high grades even when the sentiment index is mostly negative. The high value of the average grade again implies that the majority of the students were providing very high grades in their feedback while expressing negative feelings regarding the teachers and teaching tools and methods.

Given the statistical evaluation of the Student Objectivity Model towards sentiment analysis and average grade, the model indicates high rate of inconsistency in the dependency between variables: student index and average grade. This relation implies a degree of non-objectivity in students' feedback. Figure 6 presents the distribution of data-set for each defined group in Table 2 separately.

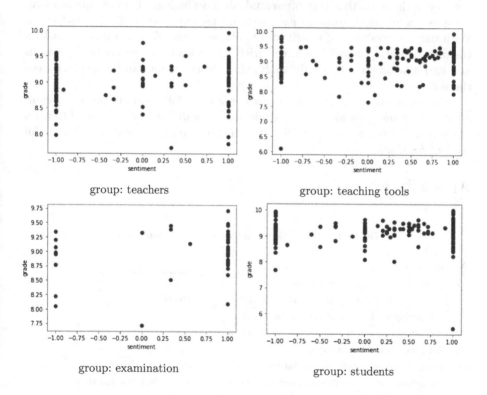

Fig. 6. Scatter plot - Student Objectivity model data-set

6   Conclusion

Educational Data Mining and Learning Analytics can contribute in the improvement of the quality of educational processes by evaluating various teaching tools and methods. Analyzing the feedback data from students is a crucial part of the process of improving learning systems, scientific-teaching methods as well as the quality of the education itself. The process will allow future generations to have better education and thus contribute to a more advanced society. In addition, students providing objective grading is of great importance to universities, because the improvement of the academic tools and methods is strictly tied to the feedback from the students and their academic success.

In this case study, a set of data from the self-evaluation system of the University was processed using several data mining tools and algorithms. Examining student satisfaction from obtained sentiment classifications showed significant correlation between comments and grades provided by the students. By analyzing the obtained results we found a high rate of inconsistencies between the emotions expressed in the free text comments and grades. We noticed an inconsistency in the sense that very often students gave high grade to the quantitative questions, while in the descriptive question they expressed their dissatisfaction. With time constraints and multitude of polls, students often just quickly click on the first answers provided. However, with strategically placed control questions and narrative answers textually analysed, a clearer result can be obtained from the polls.

The proposed model gives a correlation of the relationship between the comments and grades provided by the students. The correlation can be used to asses the validity of future polls, predict student objectivity and extract meaningful student feedback.

Appendix 1

Table 4. All questions on the self-evaluation form

No.	Questions
1	Readiness for lectures/exercises?
2	Quality of teaching process (way of presenting the material)?
3	Regularity of classes and rational use of time?
4	Availability of consultations and communication?
5	Attitude towards the student?
6	Provision of appropriate material for learning and passing the subject?
7	Compliance of the exercises (auditory, laboratory) with the lectures and their temporal coordination?
8	Usefulness of performed laboratory exercises?
9	Objectivity in the assessment and manner of realization of the exam?
10	Requirements that are offered to students (colloquia and others) and exam complexity?
11	My attendance at classes was (lectures/exercises)?
12	Leave note for improving the quality of the course

Appendix 2

Table 5. Questions and answers examples

No.	Question	Answer (Student 1)	Answer (Student 2)
1	Readiness for lectures/exercises?	9	9
2	Quality of teaching process (way of presenting the material)?	10	9
3	Regularity of classes and rational use of time?	10	10
4	Availability of consultations and communication?	10	10
5	Attitude towards the student?	10	10
6	Provision of appropriate material for learning and passing the subject?	10	8
7	Compliance of the exercises (auditory, laboratory) with the lectures and their temporal coordination?	10	10
8	Usefulness of performed laboratory exercises?	10	9
9	Objectivity in the assessment and manner of realization of the exam?	10	8
10	Requirements that are offered to students (colloquia and others) and exam complexity?	10	7
11	My attendance at classes was (lectures/exercises)?	10	10
12	Leave note for improving the quality of the course	I have no any objections	I think it takes a little more time to solve colloquium examples ...

References

1. iknow: Innovation and knowledge management towards estudent information system iknow (2019). http://iknow.ii.edu.mk/Partners.aspx?a=2
2. Abd-Elrahman, A., Andreu, M., Abbott, T.: Using text data mining techniques for understanding free-style question answers in course evaluation forms. Res. Higher Educ. J. **9**, 1 (2010)
3. Baradwaj, B.K., Pal, S.: Mining educational data to analyze students' performance. arXiv preprint arXiv:1201.3417 (2012)
4. Chowdhury, G.G.: Natural language processing. Ann. Rev. Inf. Sci. Technol. **37**(1), 51–89 (2003)
5. Daniel, B.: Big data and analytics in higher education: opportunities and challenges. Br. J. Educ. Technol. **46**(5), 904–920 (2015)
6. Devlin, J., Chang, M.W., Lee, K., Toutanova, K.: Bert: pre-training of deep bidirectional transformers for language understanding. arXiv preprint arXiv:1810.04805 (2018)

7. Kitanovski, D., Stojmenski, A., K.I., Dimitrova, V.: Implementation of novel faculty e-services for workflow automatization (2019)
8. Doleck, T., Lemay, D.J., Basnet, R.B., Bazelais, P.: Predictive analytics in education: a comparison of deep learning frameworks. Educ. Inf. Technol. **25**(3), 1951–1963 (2019). https://doi.org/10.1007/s10639-019-10068-4
9. Khaled, D.: Natural language processing and its use in education. Int. J. Adv. Comput. Sci. Appl. **5** (2014). https://doi.org/10.14569/IJACSA.2014.051210
10. Nikolovski, V., Stojanov, R., Mishkovski, I., Chorbev, I., Madjarov, G.: Educational data mining: case study for predicting student dropout in higher education (2015)
11. Pechenizkiy, M., Calders, T., Vasilyeva, E., De Bra, P.: Mining the student assessment data: lessons drawn from a small scale case study. In: Educational Data Mining 2008 (2008)
12. Sin, K., Muthu, L.: Application of big data in education data mining and learning analytics-a literature review. ICTACT J. Soft Comput. **5**(4) (2015)
13. Waheed, H., Hassan, S.U., Aljohani, N.R., Hardman, J., Alelyani, S., Nawaz, R.: Predicting academic performance of students from VLE big data using deep learning models. Comput. Hum. Behav. **104**, 106189 (2020)

Fat Tree Algebraic Formal Modelling Applied to Fog Computing

Pedro Juan Roig[1](✉) ⓘ, Salvador Alcaraz[1]ⓘ, Katja Gilly[1]ⓘ,
and Sonja Filiposka[2]ⓘ

[1] Miguel Hernández University, Elche, Spain
{proig,salcaraz,katya}@umh.es
[2] Ss. Cyril and Methodius University, Skopje, North Macedonia
sonja.filiposka@finki.ukim.mk

Abstract. Fog computing brings distributed computing resources closer to end users, thus allowing for better performance in internet of things applications. In this context, if all necessary resources worked together in an autonomous manner, there might be no need for an orchestrator to manage the whole process, as long as there is no Cloud or Edge infrastructure involved. This way, control messages would not flood the entire network and more efficiency would be achieved. In this paper, a framework composed by a string of sequential wireless relays is presented, each one being attached to a fog computing node and all of those being interconnected by a fat tree architecture. To start with, all items involved in that structure are classified into different layers, and in turn, they are modelled by using Algebra of Communicating Processes. At this stage, a couple of scenarios are being proposed: first, an ideal one where the physical path always takes the same direction and storage space is not an issue, and then, a more realistic one where the physical path may take both directions and there may be storage constraints.

Keywords: Algebraic modelling · Distributed systems · Fat tree architecture · Fog computing · Live VM migration

1 Introduction

Fog computing extends the cloud computing paradigm by bringing its capabilities closer to the end user, providing low latency, location awareness and mobility support. This way, quality of service indicators, such as latency or jitter are significantly reduced [1].

This new paradigm is catching up, to a large extend, due to the rise of internet of things (IoT), as that brings a new huge amount of devices with little computing capabilities needing to process a great deal of information [2]. Furthermore, those devices usually have poor power supplying assets, as many of them are battery-powered, hence they need to save as much energy as possible, and bringing the computing power as close to the source of information as possible helps that purpose.

© Springer Nature Switzerland AG 2020
V. Dimitrova and I. Dimitrovski (Eds.): ICT Innovations 2020, CCIS 1316, pp. 111–126, 2020.
https://doi.org/10.1007/978-3-030-62098-1_10

Fog services include not only computing power, but also other resources such as memory, storage and application services [3]. Because of all these features getting closer to the end user, fog computing may be considered as a decentralised version of cloud computing, allowing for better performance in all sort of scenarios, especially in time and location-sensitive applications.

Cloud computing may take advantage of fog computing deployments as synergies may arise between them [4]. That is possible because fog is positioned between the smart devices being managed by end users and the Cloud, hence making possible for the fog to collect raw data from end users, to process them properly and to send them up to the cloud for big data analysis and storage.

IoT is expected to grow exponentially in the coming years and the consolidation of fog computing technology is critical for its success. This will allow the introduction of brand new strategies such as a new industrial revolution called Industry 4.0, the interconnection of nearly all physical and virtual objects around us, making possible the rise of new disciplines like telemedicine or auto vehicle systems, or the emerge of new standards, such as the OpenFog Reference Architecture in order to check out system reliability and dependability.

Therefore, fog computing may be considered as a hot topic for years to come, and as such, it has to be thoroughly studied. For that reason, the target of this paper is to get a fog computing model working on a distributed basis for a particular framework made of a string of sequential wireless relays, in a way that end users may get into the topology at one end and may get out of it either at the other end or otherwise at the same end, after having spent some time inside.

Each wireless access point (AP) is connected to a fog computing asset, in a way that when a new user first gets wirelessly associated to the framework through the AP situated at one of its ends, the fog computing facility assigns a virtual machine (VM) to such a user. Then, when that user gets within wireless range of a neighbouring AP, a wireless handover occurs, and its assigned VM gets migrated to another host being as close as possible to the host directly connected to this new AP, as fat tree topology brings all fog computing sources together. And that behaviour will keep going on until the user gets out of the system through one of its ends, and then its assigned VM will be erased.

The organisation of this paper will be as follows: first, Sect. 2 introduces some Algebra of Communicating Processes (ACP) fundamentals, then, Sect. 3 presents a fat tree specification, next, Sect. 4 talks about live VM migration, in turn, Sect. 5 shows the topology framework proposed along with its building blocks, after that, Sect. 6 gets Scenario 1, where an algebraic model for physical devices going one-way through the path and without storage restrictions is shown, afterwards, Sect. 7 gets Scenario 2, where physical devices go both ways through the path and there are storage restrictions, later, Sect. 8 performs model verification and finally, Sect. 9 will draw the final conclusions.

2   Algebra of Communicating Processes

Regarding the achievement of such a model for the fog computing framework presented, formal description techniques (FDTs) are the proper way so as to get common descriptions for the same sort of actions. FDTs have been long used [5] in order to facilitate automated protocol implementation and to validate models against implementations, as well as to check out its security, such as modelling industrial applications with Petri nets or verifying security protocols with Spin. However, the FDT being probably the most useful in order to work with IT or IP protocols are mathematical methods, which are used in a wide range of fields, such as web security or wireless sensor networks.

Process algebras are mathematical languages aimed at describing and verifying the properties of concurrent and distributed communicating systems [6]. There are some process algebras, but among all them, the most abstract one is ACP, as it forgets about the real nature of objects, just focusing on the algebraic structure formed by a set of atomic actions and some operators whose properties are given by axioms [7]. Therefore, ACP may be considered as an abstract algebra such as group theory, ring theory or field theory, meaning that protocol verification may be done by using proof by mathematical induction or proof by contradiction.

Thanks to that approach as an abstract algebra, this is, processes being conceived as algebraic equations, it is possible to employ purely algebraic structures to reason about them so as to deal with processes. That fact allows to model a real process with some ACP process terms being behaviourally equivalent as that process being modelled, which is known as bisimilarity or bisimulation equivalence [8]. Focusing on the models proposed, ACP atomic actions have been used, such as send or read a message through a channel z, namely s_z or r_z, as well as some basic operators to describe their relationships, such as sequential ".", alternate "+", concurrent "||" or conditional (True ◁ $condition$ ▷ False).

3   Fat Tree Network Architecture

Data centers may contain thousands of servers requiring a significant aggregate bandwidth and having remarkable processing capabilities. In order to achieve those goals, many of the schemes applied are based on Clos networks, a type of multistage circuit switching network originally designed for traditional plain old telephone system circuits delivering high bandwidth, where connections among a great deal of input and output ports may be established by means of only small-sized switches [9]. In particular, special instances of that topology have been adapted in order to fit the increasing need for more flexible and scalable schemes. There are many variations around, most of them using commodity ethernet layer-3 switches so as to produce scalable data centers, but their approach differs when dealing with a trade-off among the number of connections allowed, latency predictability and scalability [10].

With respect to fat tree, it is a multi-tier Clos network, although the most common design is a 3-tier scheme. Sticking to that case [11], a k-ary fat tree consists of some k-port switches along with compute nodes like servers attached to the lower layer. Its basic building block is named a pod and there are k pods, each one containing two layers of $k/2$ switches, namely edge layer the lower one and aggregation layer the upper one. Within a pod, each switch in the edge layer is connected to $k/2$ compute nodes, whilst each of the other $k/2$ ports is connected to $k/2$ of the k ports in the aggregation layer. Also, there are $(k/2)^2$ switches in the core layer, each one being connected to each of the k pods.

Therefore, the network design for fat tree is not full-mesh but a partial-mesh hierarchical one. This architecture improves traffic management and scalability because the number of connections gets bounded as network expands, but it gets penalised when it comes to internal traffic as the distance among servers depends on their location within the topology. In addition to this, oversubscription [12] is defined as the possibility of one particular host to communicate with any other given host within the infrastructure at the full bandwidth of the network interface, being that case stated by the ratio 1:1, which is the most referenced in literature, although it might raise costs.

4   Live VM Migration

Fat tree topologies have been widely used in cloud computing paradigm, and its use may well be expanded to fog computing, as it may be considered as an extension of the former services to the edge of the network. A key concept in both environments is virtualisation, enabling VMs to coexist in a physical server (host) to share its resources. Those VMs may be associated to users, and as those users move around, VM migration may take place among physical servers in order for VMs to be as close as possible to their associated users [13]. This way, VM placement and migration provides an optimised usage of available resources, resulting in both energy efficiency and performance effectiveness [14].

In order to undertake the live VM migration process between two hosts, some approaches have been proposed, although a pre-copy migration process seems to be the predominant one [15]. To start with, a target host is selected and necessary resources therein are reserved in advanced for the source VM to be migrated. Then, an iterative pre-copy stage is performed, where RAM content of the running VM is moved from source host to target host in the first round, and in turn, only modified pages, also known as dirty pages, are moved in subsequent iterations until the dirty pages are small enough. At that point, source VM is halted in order to copy CPU state and remaining inconsistencies to target VM, resulting in two consistent copies of the same VM in both source and target host. Eventually, source VM gets discarded and target VM is activated, therefore, live VM migration has been successfully completed.

5   Topology Framework

After having presented the key concepts, it is time to introduce the working framework. This is a fog computing environment for IoT devices moving around within a wireless domain, in such a way that, as they move along, their associated computing assets are to follow that movement. That framework may be divided into layers, such as the wireless layer for the physical IoT to move about, the host layer for the associated VM to each IoT to be hosted and the fat tree layers for the live VM migration to take place, preferring that topology over others like leaf and spine in order to avoid scalability issues, as explained before.

To start with, the *wireless layer* is composed by a string of sequential wireless relays, in a way that one is reached right after the previous one and just before the following one, making a well-ordered set in the same way as the set of natural numbers. This disposition might be applied to an array of sequential wireless nodes, placed in an ordered manner in any type of linear structure such as a railway lines, roads or pipelines. In this kind of layout, IoT devices acting as end users may enter the topology only at one end and may leave it just at either end but not at any intermediate node.

Each of those wireless nodes has a fog computing asset associated to it, namely a VM hosted at one of the servers being part of a fog computing facility situated at the *host layer*. At the time a new end user gets connected for first time into the system through one of the wireless nodes at the ends, a new VM is to be created and associated to such a user in a host, and two options have been considered herein. The first one is to create the VM in the host directly connected to that wireless node, thus, it will be done only if there are available resources left in that particular host, otherwise, the VM is not going to be created and, as a consequence, the end user is going to be ejected out of the system. As per the second one, after trying to allocate VM in its associated host unsuccessfully, VM allocation may take place in another host, preferably belonging to the same pod, as long as any of those have enough resources left.

Once into the system, when an existing user moves to a neighbouring wireless node, this is, when that user gets within a good wireless range of another node, a wireless handover occurs, and therefore, its associated VM tries to move to the host directly connected to that new node. This migration will be completed whether there are available resources in that new host, otherwise, migration will not take place, so that VM will stay in the same host as it was previously. Eventually, after a string of wireless handovers by an end user and its corresponding attempts of live VM migration, the end user will get out of the system through one of its ends, which will cause its associated VM to be erased, and consequently, its computing resources will be liberated again.

Regarding the fat tree infrastructure interconnecting all fog computing facilities, a three-tier scheme with a generic k value and oversubscription ratio $1 : 1$ is going to be deployed, hence there are three more layers to be mentioned, corresponding to each layer defined in that fat tree topology, this is, *edge layer*, *aggregation layer* and *core layer*. The first one is composed by the switches being directly connected to the hosts, this is, fog computing facilities, the second one contains the switches interconnecting devices within a single pod, and the third one involves all switches interconnecting the different pods. This structure allows

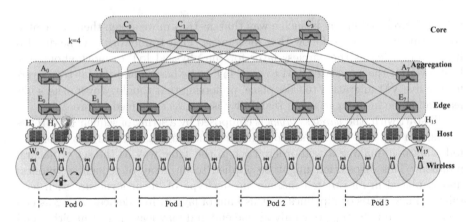

Fig. 1. Topology (if $k = 4$)

a total amount of $k^3/4$ hosts, and therefore, that number of wireless nodes will be directly connected to them, one to each host.

In summary, Fig. 1 shows this working framework, where an IoT device is standing at W_1 and its associated VM is at H_1. The next possible moves for that end user are either going to the left to W_0 or going to the right to W_2, that triggering the process of live VM migration to get its VM the closest possible.

With respect to identification of the different devices within this topology, this is done on a layer basis, this is, with the initial letter of the corresponding layer in capital letters, followed by a natural number indicating its order in that layer, starting from zero and moving from left to right. As per the layers, let us remember that there are five of them: Wireless *(W)*, Host *(H)*, Edge *(E)*, Aggregation *(A)* and Core *(C)*. In addition to it, as per the device number within a single layer, the total amount N of devices per layer are: wireless $(k^3/4)$, hosts $(k^3/4)$, edge $(k^2/2)$, aggregation $(k^2/2)$, core $(k^2/4)$, hence, the devices will be assigned a natural number going from 0 onwards up to $N - 1$, in order to take advantage of congruence relation in modular arithmetic [16]. The modulo-m operator is usually expressed by *mod* but it will be herein expressed by $|_m$ as it looks more compact.

With regard to connections between any two devices, each type is identified in a similar way, but with a different nomenclature. Connections between two wireless layer nodes are called paths, hence, they are named after its initial letter p in lower letters, followed by the number corresponding to the node being the right end of the path. That makes p_0 for the path coming from outside into the system from the left, and $p_{k^3/4}$ for the path coming in from the right. This way, a generic W_i may reach its neighbour W_{i+1} through path p_{i+1}, whereas its neighbour W_{i-1} through path p_i. With reference to the rest of connections, they are identified by the port numbers of their corresponding ends. Figure 2 shows the port schemes for each sort of device and the layout of their lower and upper ports.

As per the behaviour of each kind of switch, if an edge receives a message from a lower port, it will forward it to the destination host if both source and destination hang on that same switch, otherwise, it will send it over through all upper ports, whereas if a message is received from an upper port, it will forward it to the destination host. With respect to the aggregation switches, an incoming message through a lower port will be forwarded to the edge where destination host is hanging on if both source and destination share the same pod, otherwise, it will send it over through all upper ports, whereas if a message is delivered from an upper port, it will forward it to the edge connecting to the destination host. With regard to the core switches, when they receive a message through a port, they will forward it to the pod where destination host is located.

Putting together all devices and connections, two scenarios are going to be studied by undertaking an algebraic formal description using ACP syntax and semantics for a single user, which may easily be extended to n users, as shown below. First off, an ideal scenario where physical paths always take the same direction, namely coming from left to right, and storage space is not an issue, thus supposing there will always be available resources for a new VM. Then, a more realistic scenario where physical paths may take both directions and resources are limited, hence having restrictions for hosting a new VM.

Finally, each scenario will have its own particular expressions in order to describe the behaviour of just one moving device. But in both instances, many devices may get into the system in a concurrent manner, therefore the overall algebraic expression in both cases, may be easily extended for n users:

$$X = \left\|\begin{matrix}n\\ \\u=0\end{matrix}\right. \left(\left\|_{i=0}^{\frac{k^3}{4}-1} \right\|_{i'=0}^{\frac{k^3}{4}-1} \left\|_{j=0}^{\frac{k^2}{2}-1} \right\|_{j'=0}^{\frac{k^2}{2}-1} \left\|_{l=0}^{k-1} \left(W_i^u \| H_{i'}^u \| E_j^u \| A_{j'}^u \| C_l^u \right) \right) \cdot X$$

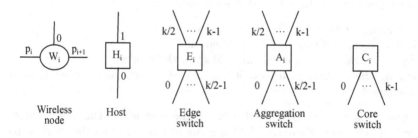

Fig. 2. Connection schemes

6 Scenario 1: One-Way and No Resource Restrictions

This first scenario represents physical devices going just one way, namely from left to right. In other words, IoT devices getting into the system from wireless node 0, going all the way through the different wireless nodes in a sequential manner and finally getting out of the system out of wireless node $k^3/4$. Additionally, there are no storage constraints, meaning that there will always be available resources to either create a new VM in a host to be associated with any new incoming device, or otherwise to perform a live VM migration to any host.

The sequence of events for this case scenario is listed below, and after that, its counterpart algebraic expressions using ACP are presented:

1. Initially, a new IoT device without VM comes into W_0: first, it gets associated with AP_0, then, it requests a new VM to H_0, called FOG_0, and upon confirmation, it gets bound to it; eventually, at a later stage, it moves to W_1.
2. At some point, that IoT device moves into a neighbour W_i from W_{i-1}: first, it gets disassociated with AP_{i-1} and associated with AP_i, then, it requests its VM to be moved from the host it was on (in this case scenario is always $i - 1$) to the host directly connected to its current wireless node (i), with variables x and y holding generic destinations one way and the other, hence in this case, VM gets unbound from H_{i-1} and then it gets bound to H_i according to the pre-copy live VM migration strategy supposing just one iteration, this is, requesting a copy of the source VM at destination, and in turn, discarding the old one and activating the new one upon confirmation; eventually, at a later time, the IoT will move to W_{i+1} and this will go again.
3. Finally, that IoT device will leave the system after reaching $W_{\frac{k^3}{4}-1}$: upon reassociating with $AP_{\frac{k^3}{4}-1}$ and getting rebound to host $H_{\frac{k^3}{4}-1}$, it will eventually go out of the system at a certain point, but before that, the IoT devices gets unbound from that host, then, its VM is requested to be erased, and in turn, and it gets disassociated from that AP.

Messages used for Live VM Migration are four:

- requesting an existing VM to the source
- sending a copy of that VM to destination
- requesting the erasement of that VM from the source
- sending the confirmation to destination and in turn, upon receipt, activate that new VM

Those messages are exchanged between the host where the VM is currently located and the host where that VM is going to move, taking the shortest paths between them both through the fat tree infrastructure. Regarding the direction of the flow, the new host sends a request message for a copy of the VM to the old host, which in turn, replies back with a copy message, and upon receipt, a new copy is inserted into the new host. After that, an erase message is sent from the new host to the old one, and upon receipt, the VM is deleted from the old host, which in turn, sends back a done message, confirming the success of the migration, which in turn, activates the new VM.

In order to keep things simple, all four messages are going to be expressed by the generic notation $MSG(x,y)$, where x and y are the source and destination of the message being carried out through the fat tree infrastructure, whilst i represents the host attached to the wireless node where the moving item is connected. Additionally, a and b are the source and destination hosts for a migration, and it is to be noted that the condition for two hosts to be hanging on the same edge switch is $\left\lfloor \frac{a}{k/2} \right\rfloor = \left\lfloor \frac{b}{k/2} \right\rfloor$, whereas the condition for two hosts to be connected to the same pod is $\left\lfloor \frac{a}{(k/2)^2} \right\rfloor = \left\lfloor \frac{b}{(k/2)^2} \right\rfloor$, whilst switch ports (q) go from 0 to $k-1$.

Here they are the algebraic expressions for all five entities in scenario 1:

$$W_i = \sum_{i=0}^{\frac{k^3}{4}-1} ((r_{p_0}(IoT(-)) \cdot AP_0 \cdot s_0(INIT) \cdot r_0(0) \cdot BIND(IoT(0)) \cdot s_{p_1}(IoT(0)) \cdot W_1)$$

$$\lhd\, i = 0 \,\rhd$$

$$(r_{p_i}(IoT(i-1)) \cdot \neg AP_{i-1} \cdot AP_i \cdot s_0(MOVE(i-1,i)) \cdot r_0(i) \cdot UNBIND(IoT(i-1))$$

$$\cdot\, BIND(IoT(i))$$

$$\cdot\, ((UNBIND(IoT(i)) \cdot s_0(KILL) \cdot r_0(ACK) \cdot \neg AP_{\frac{k^3}{4}-1} \cdot s_{P_{\frac{k^3}{4}}}(IoT(-)))$$

$$\lhd\, i = \frac{k^3}{4} - 1 \,\rhd$$

$$(s_{p_{i+1}}(IoT(i)) \cdot W_{i+1}))))))$$

$$H_i = \sum_{i=0}^{\frac{k^3}{4}-1} ((r_0(INIT) \cdot FOG_0 \cdot s_0(0)$$

$$\lhd\, i = 0 \,\rhd$$

$$(r_0(KILL) \cdot \neg FOG_{\frac{k^3}{4}-1} \cdot s_0(ACK)$$

$$\lhd\, i = \frac{k^3}{4} - 1 \,\rhd$$

$$((r_0(MOVE(i-1,i)) \cdot s_1(RQT(i,i-1))) + (r_1(RQT(i+1,i)) \cdot s_1(COPY(i,i+1)))$$

$$+ (r_1(COPY(i-1,i)) \cdot FOG_i \cdot s_1(ERASE(i,i-1))) + (r_1(ERASE(i+1,i)) \cdot \neg FOG_i$$

$$\cdot\, s_1(DONE(i,i+1))) + (r_1(DONE(i-1,i)) \cdot s_0(i))))))) \cdot H_i$$

$$E_j = \sum_{j=0}^{\frac{k^2}{2}-1} (\sum_{q=0}^{\frac{k}{2}-1} r_q(MSG(x,y)) \cdot (s_{b_{\lfloor k/2 \rfloor}}(MSG(x,y)) \lhd \left\lfloor \frac{a}{k/2} \right\rfloor = \left\lfloor \frac{b}{k/2} \right\rfloor \rhd \sum_{q=\frac{k}{2}}^{k-1} s_q(MSG(x,y)))$$

$$+ \sum_{q=\frac{k}{2}}^{k-1} (r_q(MSG(x,y)) \cdot s_{b_{\lfloor k/2 \rfloor}}(MSG(x,y))) \cdot E_j$$

$$A_j = \sum_{j=0}^{\frac{k^2}{2}-1} (\sum_{q=0}^{\frac{k}{2}-1} r_q(MSG(x,y)) \cdot (s_{\left\lfloor \frac{b}{k/2} \right\rfloor}(MSG(x,y)) \lhd \left\lfloor \frac{a}{(k/2)^2} \right\rfloor = \left\lfloor \frac{b}{(k/2)^2} \right\rfloor \rhd \sum_{q=\frac{k}{2}}^{k-1} s_q(MSG(x,y)))$$

$$+ \sum_{q=\frac{k}{2}}^{k-1} (r_q(MSG(x,y)) \cdot s_{\left\lfloor \frac{b}{k/2} \right\rfloor}(MSG(x,y)))) \cdot A_j$$

$$C_l = \sum_{l=0}^{\frac{k^2}{4}-1} (\sum_{q=0}^{k-1} r_q(MSG(x,y)) \cdot (s_{\left\lfloor \frac{b}{(k/2)^2} \right\rfloor}(MSG(x,y)))) \cdot C_l$$

7 Scenario 2: Two-Way and Resource Restrictions

This second scenario represents physical devices going both ways, namely, getting in or out of the system at any both ends, as well as moving around at any given time. Therefore, IoT devices may enter into the system or leave it at wireless nodes 0 or $k^3/4$, and may change the sense of direction at random. In addition to it, there are storage constraints, that being closer to real deployments, in a way that generation and migration of VMs may be undertaken by another host located anywhere, in case the host directly connected to a particular relevant wireless node has not available resources to do it, or in the worst case scenario, VMs might be neither created nor migrated if no resources are at all available within the whole pod.

Taking the previous model as a base, there have been introduced some changes in order to reflect the features of this new scenario:

- Path directions may be distinguished, thus, r_{p_i} will mean taking Path i from left to right, whereas $r_{p'_i}$ will mean doing it the other way around, hence the model for W_i needs to be adjusted to reflect each path direction, taking into account the moving item may change direction at any time.
- VMs are located in any host, therefore, VMs will be considered to be in a generic location g, and the binding of an IoT device with its associated VM will be denoted by $IoT(g)$. Furthermore, if a VM is not able to be created due to lack of available resources anywhere, it is reflected by using NAK.
- VM operations, established from host to host through the fat tree structure, may include a new argument at the beginning called A, standing for Action, indicating whether it is about Initializing (I), Killing (K) or Moving (M). Therefore, generic messages will have the structure $MSG(A, x, y)$, extending the one explained in scenario 1.

In order to deal with the search for another host, either for initialising or moving, it is necessary to establish a general strategy. First off, the host directly connected to the wireless node where the moving item is associated at a given time, that being the ideal case. If such a host is not available to allow a new VM inside, the hosts connected to the same edge switch will be tested, then, the hosts connected to the same pod will be scanned, and eventually, the hosts elsewhere within the topology will be looked up.

Regarding each sort of VM control flow, these are the corresponding messages taking part on each of them and their meaning:

- Initialising a VM ($INIT(i)$): First off, i host is checked out for available resources, and if it does have, then the VM is created over there. However, if this is not the case, query messages $QRY(i,d)$ are sent to look for available resources in other hosts d, according to the order proposed before. The answer for those queries are Response messages $RSP(d,i)$, each of them having two possible answers, such as $FREE$ or $FAIL$. In case of the former response at any point, indicating that distant host d meets the requirements, an $ASSIGN(i,d)$

message will be issued in order to create the proper VM in the remote host, being replied back with a $DONE$(d,i) message. In case of the latter response by all hosts, a NAK message is sent to the wireless node, meaning that there are no available resources for new VMs in the system, hence, the moving device is kicked off. However, if that VM has been created in any host, its generic host identifier g is forwarded on to the wireless node for the IoT device to get bound to it, where g means the host where the VM has been created, either i or d.

- Migrating a VM ($MOVE(f, i)$): The process to search for the closest host with available resources to house a VM is pretty similar than the one described above, by first checking out the host i looking for available resources to carry out the migration, and if that is not the case, by then exchanging QRY(i,d) and RSP(d,i) messages looking for the closest host d with those available resources. In that way, if a $FREE$ response is obtained at any point, then a comparison is executed in order to check whether the available host d is closer from i than the host currently keeping the VM, namely host f, regarding the number of hops away from host i. This action will be regarded by comparing the hop count between i and f with the hop count between i and d, this is, $HOPS_{i-f}$ and $HOPS_{i-d}$, and only if the new hop distance is lower, then VM migration will take place between host f and host d. On the one hand, if the VM migration is decided, its related message exchange is the same as explained in the previous scenario, adding up a new parameter related to host i to indicate the start and get the acknowledge receipt at the end of the whole process. Those messages are $RQT_i(d, f)$, $COPY_i(f, d)$, $ERASE_i(d, f)$ and $DONE_i(f, d)$, and if neither f nor d are the local host i, a further message $RELAY_i(d, i)$ is used to confirm back the operation successfully. Eventually, the host identifier of the new host holding the VM is sent to the wireless node for the IoT device to get bound to it, in the form of generic host g, including the cases where i or d are the new VM destination. On the other hand, if VM migration is not decided, a message with the host f is sent to the wireless node, meaning that VM migration is not going to be performed, thus VM keeps on the same host as it is allocated.

- Killing a VM ($KILL(g)$): This process just deletes a VM standing on generic host g, corresponding to a moving device which is about to leave the system. This action is performed by means of two messages, such as $DELETE(i, g)$ and $DONE(g, i)$, and after that, an ACK message is sent to the wireless node i, causing the moving device to get dissassociate from the AP where it is connected to, and in turn, it gets off the system.

Here they are the algebraic expressions for all five entities in scenario 2:

$$C_l = \sum_{i=0}^{\frac{k^2}{4}-1} (\sum_{q=0}^{k-1} r_q(MSG(A, x, y)) \cdot (s_{\left\lfloor \frac{b}{(k/2)^2} \right\rfloor}(MSG(A, x, y)))) \cdot C_l$$

$$E_j = \sum_{j=0}^{\frac{k^2}{2}-1} (\sum_{q=0}^{\frac{k}{2}-1} r_q(MSG(A,x,y)) \cdot (s_{b_{|k/2}}(MSG(A,x,y)) \lhd \left\lfloor \frac{a}{k/2} \right\rfloor = \left\lfloor \frac{b}{k/2} \right\rfloor \rhd \sum_{q=\frac{k}{2}}^{k-1} s_q(MSG(A,x,y)))$$

$$+ \sum_{q=\frac{k}{2}}^{k-1} (r_q(MSG(A,x,y)) \cdot s_{b_{|k/2}}(MSG(A,x,y))) \cdot E_j$$

$$H_i = \sum_{i=0}^{\frac{k^3}{4}-1} ((r_0(INIT(i))$$

$$\cdot (FOG_i \cdot s_0(i)$$

$$\lhd FREE_i \rhd$$

$$\sum_{d=j \cdot k/2}^{((j+1) \cdot (k/2)-1)-\{i\}} (s_1(QRY(I,i,d)) \cdot (s_1(ASSIGN(I,i,d)) + r_1(DONE(I,d,i)) \cdot s_0(d)$$

$$\lhd RSP(I,d,i) = FREE_d \rhd$$

$$\sum_{d=j \cdot (k/2)^2}^{((j+1) \cdot (k/2)^2-1)-\{i\}} (s_1(QRY(I,i,d)) \cdot (s_1(ASSIGN(I,i,d)) + r_1(DONE(I,d,i)) \cdot s_0(d)$$

$$\lhd RSP(I,d,i) = FREE_d \rhd$$

$$\sum_{d=0}^{((k^3/4)-1)-\{i\}} (s_1(QRY(I,i,d)) \cdot (s_1(ASSIGN(I,i,d)) + r_1(DONE(I,d,i)) \cdot s_0(d)$$

$$\lhd RSP(I,d,i) = FREE_d \rhd$$

$$s_0(NAK)))))))$$

$$+ r_1(ASSIGN(I,d,i)) \cdot FOG_i \cdot s_1(DONE(I,i,d)))$$

$$+ (r_0(KILL(g))$$

$$\cdot (\neg FOG_i \cdot s_0(ACK)$$

$$\lhd i = g \rhd$$

$$s_1(DELETE(K,i,g)) \cdot r_1(DONE(K,g,i)) \cdot s_0(ACK))$$

$$+ r_1(DELETE(K,g,i)) \cdot \neg FOG_i \cdot s_1(DONE(K,i,g)))$$

$$+ ((r_0(MOVE(f,i))$$

$$\cdot (s_0(i)$$

$$\lhd FOG_i \rhd$$

$$(s_1(RQT_i(M,i,f)) + r_1(COPY_i(M,f,i)) \cdot FOG_i \cdot s_1(ERASE_i(M,i,f))$$

$$+ r_1(DONE_i(M,f,i)) \cdot s_0(i)$$

$$\lhd FREE_i \rhd$$

$$\sum_{d=j \cdot k/2}^{((j+1) \cdot (k/2)-1)-\{i\}} ((s_1(QRY(M,i,d)) \cdot (s_1(RQT_i(M,d,f)) + r_1(COPY_i(M,f,d))$$

$$\cdot FOG_d \cdot s_1(ERASE_i(M,d,f)) + r_1(DONE_i(M,f,d)) \cdot s_1(RELAY_i(M,d,i))$$

$$\lhd HOPS_{i,d} < HOPS_{i,f} \rhd$$

$$s_0(f))$$

$$\lhd RSP(M,d,i) = FREE_d \rhd$$

$$\sum_{d=j\cdot(k/2)^2}^{((j+1)\cdot(k/2)^2-1)-\{i\}} ((s_1(QRY(M,i,d))\cdot(s_1(RQT_i(M,d,f))+r_1(COPY_i(M,f,d))$$

$$\cdot FOG_d\cdot s_1(ERASE_i(M,d,f))+r_1(DONE_i(M,f,d))\cdot s_1(RELAY_i(M,d,i))$$

$$\lhd\ HOPS_{i,d}<HOPS_{i,f}\ \rhd$$

$$s_0(f))$$

$$\lhd\ RSP(M,d,i)=FREE_d\ \rhd$$

$$s_0(f))))))$$

$$+r_1(RQT_i(M,f,i))\cdot s_1(COPY_i(M,i,f))$$

$$+r_1(ERASE_i(M,f,i))\cdot\neg FOG_i\cdot s_1(DONE_i(M,i,f))$$

$$+r_1(RQT_i(M,d,f))\cdot s_1(COPY_i(M,f,d))$$

$$+r_1(ERASE_i(M,d,f))\cdot\neg FOG_f\cdot s_1(DONE_i(M,f,d))$$

$$+r_1(RELAY_i(M,d,i))\cdot s_0(d)))\cdot H_i$$

$$A_j=\sum_{j=0}^{\frac{k^2}{2}-1}(\sum_{q=0}^{\frac{k}{2}-1}r_q(MSG(A,x,y))\cdot(s_{\lfloor\frac{b}{k/2}\rfloor}(MSG(A,x,y))\lhd\left\lfloor\frac{a}{(k/2)^2}\right\rfloor=\left\lfloor\frac{b}{(k/2)^2}\right\rfloor\rhd\sum_{q=\frac{k}{2}}^{k-1}s_q(MSG(A,x,y)))$$

$$+\sum_{q=\frac{k}{2}}^{k-1}(r_q(MSG(A,x,y))\cdot s_{\lfloor\frac{b}{k/2}\rfloor}(MSG(A,x,y))))\cdot A_j$$

$$W_i=\sum_{i=0}^{\frac{k^3}{4}-1}(((r_{p_0}(IoT(-))\cdot AP_0\cdot s_0(INIT(i))\cdot r_0(g)\cdot BIND(IoT(g))$$

$$\cdot(s_{p_1}(IoT(g))\cdot W_1+UNBIND(IoT(g))\cdot s_0(KILL(g))\cdot r_0(ACK)$$

$$\cdot\neg AP_0\cdot s_{p'_0}(IoT(-))+r_0(NAK)\cdot\neg AP_0\cdot r_{p'_0}(IoT(-))))$$

$$\lhd\ i=0\ \rhd$$

$$(r_{p'_{\frac{k^3}{4}-1}}(IoT(-))\cdot AP_{\frac{k^3}{4}-1}\cdot s_0(INIT(i))\cdot r_0(g)\cdot BIND(IoT(g))$$

$$\cdot(s_{p'_{\frac{k^3}{4}-1}}(IoT(g))\cdot W_{\frac{k^3}{4}-2}+UNBIND(IoT(g))\cdot s_0(KILL(g))\cdot r_0(ACK)$$

$$\cdot\neg AP_{\frac{k^3}{4}-1}\cdot s_{p_{\frac{k^3}{4}}}(IoT(-))+r_0(NAK)\cdot\neg AP_{\frac{k^3}{4}-1}\cdot r_{p_{\frac{k^3}{4}}}(IoT(-))))$$

$$\lhd\ i=\frac{k^3}{4}-1\ \rhd$$

$$((r_{p_i}(IoT(g))\cdot\neg AP_{i-1}\cdot AP_i\cdot s_0(MOVE(g,i))\cdot(r_i(g')\cdot UNBIND(IoT(g))$$

$$\cdot BIND(IoT(g'))+r_0(g))\cdot((UNBIND(IoT(g))\cdot s_0(KILL(g))\cdot r_0(ACK)$$

$$\cdot\neg AP_{\frac{k^3}{4}-1}\cdot s_{p_{\frac{k^3}{4}}}(IoT(-))+s_{p'_i}(IoT(g))\cdot W_{i-1})$$

$$\lhd\ i=\frac{k^3}{4}-1\ \rhd$$

$$(s_{p_{i+1}}(IoT(g))\cdot W_{i+1}+s_{p'_i}(IoT(g))\cdot W_{i-1})))$$

$$+(r_{p'_i+1}(IoT(g))\cdot\neg AP_{i+1}\cdot AP_i\cdot s_0(MOVE(g,i))\cdot(r_i(g')\cdot UNBIND(IoT(g))$$

$$\cdot BIND(IoT(g'))+r_0(g))\cdot((UNBIND(IoT(g))\cdot s_0(KILL(g))\cdot r_0(ACK)$$

$$\cdot\neg AP_0\cdot s_{p'_0}(IoT(-))+s_{p_{i+1}}(IoT(g))\cdot W_{i+1})$$

$$\lhd\ i=0\ \rhd$$

$$(s_{p'_i}(IoT(g))\cdot W_{i-1}+s_{p_{i+1}}(IoT(g))\cdot W_{i+1})))))$$

8   Verification

To start with, as per the sequence of wireless relays proposed, numbered from 0 to $\frac{k^3}{4} - 1$, is clearly order isomorphic to an initial segment of natural numbers ordered by $<$, meaning that total order is established herein as antisymmetry, transitivy and totality properties apply. With that in mind, and considering that ACP is an abstract algebra, verification may be undertaken using proof by mathematical induction or proof by contradiction.

Let us consider what happen with a moving item getting into the system. Taking the first case scenario, proof by induction it is fulfilled as it is obvious that after visiting W_n, a moving item will go to W_{n+1}, hence, an item getting into the system from W_0 will always leave it out of $W_{\frac{k^3}{4}-1}$. On the other hand, proof by contradiction is also met as it does not make any sense that a moving item is forever into the system, as it will eventually reach the system exit point. Regarding the second case scenario, the same proofs apply, although it might take longer to get out of the system, as a moving item getting into the system at any end will be moving randomly, but it will at some point reach any end and eventually leave the system after a certain number of movements.

Let us now consider what happen with a VM created when a moving item first enters the system. As per the first scenario, that VM will be following the item on its way throughout the system, visiting the hosts directly connected to the wireless nodes that the item is associated to at any time, as there are no restrictions when proceeding to live VM migration, until the item is to leave the system, and at that point, that VM will be killed. As per the second scenario, the VM might be allocated in a different host that the one directly connected to the wireless node the item is connected to, but at each move, the system will try and get to allocate that VM as close as possible to its associated item, but eventually the VM gets killed when the item gets out of the system.

This is the way it goes because all channels deliver communication, as for each message sent at one end, there is a message read for that same content at the other end, whereas there is deadlock if content differs. Additionally, the rest of actions are internal, thus silent actions regarding the system external behaviour, hence send and read are the only relevant actions regarding internal system behaviour. Finally, focusing just on the external behaviour, it happens that any item getting into the system will eventually get out of it, therefore, the system might be seen as a black box, with different entry and exit points in scenario 1, although both points may or may not be the same in scenario 2.

9   Conclusions

In this paper, a new framework for providing fog computing services to IoT devices in a sequential wireless node environment has been proposed, ideal in linear schemes such as any kind of roads, railways or pipelines. Two case scenarios have been described herein, first, an ideal one with just one way path through the sequential wireless structure and unlimited computing resources when dealing with VM, thus having a VM on the host directly connected, and then, a more realistic one with a two way path and limited computing resources, as well as the option of having a VM on any host throughout the infrastructure. Algebraic models have been developed for both models and further verification has been undertaken, resulting in the fulfillment of the requirements in both cases, meaning that both algebraic models have been verified.

References

1. Yousefpour, A., Ishigaki, G., Jue, J.P.: Fog computing: towards minimizing delay in the Internet of Things. In: Narhstedt, K., Zhu, H. (eds.) 2017 IEEE 1st International Conference on Edge Computing, pp. 17–24. https://doi.org/10.1109/IEEE.EDGE.2017.12
2. Iorga, M. et al.: Fog Computing Conceptual Model - Recommendations of the National Institute of Standards and Technology. US Department of Commerce (2018). https://doi.org/10.6028/NIST.SP.500-325
3. Khan, S., Parkinson, S., Qin, Y.: Fog computing security: a review of current applications and security solutions. J. Cloud Comput. **6**(1), 1–22 (2017). https://doi.org/10.1186/s13677-017-0090-3
4. Stojmenovic, I.: Fog computing: a cloud to the ground support for smart things and machine-to-machine networks. In: Gregory, M. (eds.) Australasian Telecommunication Networks and Applications Conference, pp. 117–122. ATNAC, Melbourne (2014). https://doi.org/10.1109/ATNAC.2014.7020884
5. Quemada, J.: Formal description techniques and software engineering: some reflections after 2 decades of research. In: de Frutos-Escrig, D., Núñez, M. (eds.) Formal Techniques for Networked and Distributed Systems FORTE 2004, vol. 3235, pp. 33–42. Springer, Heidelberg (2004). https://doi.org/10.1007/978-3-540-30232-2
6. Padua, D.: Encyclopedia of Parallel Computing. Springer, Heidelberg (2011). https://doi.org/10.1007/978-0-387-09766-4
7. Lockefeer, L., Williams, D.M., Fokkink, W.: Specification and verification of TCP extended with the window scale option. In: Lang, F., Flammini, F. (eds.) Formal Methods for Industrial Critical Systems 2014, Science of Computer Programming, vol. 118, pp. 3–23. Elvesier, Amsterdam (2016). https://doi.org/10.1016/j.scico.2015.08.005
8. Fokkink, W.: Introduction to Process Algebra. Springer, Heidelberg (2000). https://doi.org/10.1007/978-3-662-04293-9
9. Jyothi, S.A., Dong, M., Godfrey, P.B.: Towards a flexible data center fabric with source routing. In: Proceedings of the 1st ACM SIGCOMM Symposium on Software Defined Networking Research, Article No. 10. ACM, New York (2015). https://doi.org/10.1145/2774993.2775005
10. Adda, M., Peratikou, A.: Routing and fault tolerance in Z-Fat tree. IEEE Trans. Parallel Distribut. Syst. **28**(8), 2373–2386 (2017). https://doi.org/10.1109/TPDS.2017.2666807
11. Al-Fares, M., Loukissas, A., Vahdat, A.: A scalable, commodity data center network architecture. ACM SIGCOMM Comput. Commun. Rev. **38**(4), 63–74 (2008). https://doi.org/10.1145/1402946.1402967

12. Guo, Z., Duan, J., Yang, Y.: Oversubscription bounded multicast scheduling in fat-tree data center networks. In: 2013 IEEE 27th International Symposium on Parallel and Distributed Processing (IPDPS), pp. 598–600. IEEE (2013). https:// doi.org/10.1109/IPDPS.2013.30
13. Kaur, P., Rani, A.: Virtual machine migration in cloud computing. Int. J. Grid Distrib. Comput. 8(5), 337–342 (2015). https://doi.org/10.14257/ijgdc.2015.8.5. 33
14. Filiposka, S., Mishev, A., Juiz, C.: Community-based VM placement framework. J. Supercomputing 71(12), 4504–4528 (2015). https://doi.org/10.1007/s11227-015- 1546-1
15. Osanaiye, O., Chen, S., Yan, Z., Lu, R., Choo, K.R., Dlodlo, M.: From cloud to fog computing: a review and a conceptual live VM migration framework. IEEE Access 5, 8284–8300 (2017). https://doi.org/10.1109/ACCESS.2017.2692960
16. Roig, P.J., Alcaraz, S., Gilly, K., Juiz, C.: Modelling VM migration in a fog computing environment. Elektronika Ir Elektrotechnika 25(5), 75–81 (2019). https:// doi.org/10.5755/j01.eie.25.5.24360

A Circuit for Flushing Instructions from Reservation Stations in Microprocessors

Dejan Spasov[✉]

Faculty of Computer Science and Engineering, Skopje, Republic of North Macedonia
dejan.spasov@finki.ukim.mk

Abstract. Modern processors include reservation stations to host instructions that are waiting to be sent to the execution units. Instructions in reservation stations may be waiting for source operands to become ready. An instruction may be executed with an exception. Responsive to the exception event, instructions younger than the instruction executed with exception need to be flushed from the reservation stations. We propose a circuit for flushing instructions from reservation stations. The proposed circuit is based on wrap bits and reorder buffer indexes to determine relative age between instructions. Wrap bit and reorder buffer index assigned to the instruction executed with exception are compared with wrap bits and reorder buffer indexes assigned to instructions in reservation stations. Instructions younger than the instruction executed with exception are flushed from the reservation stations. We compare the propagation delay and hardware complexity of the proposed circuit with the conventional flush logic based on branch masks.

Keywords: Superscalar · Reorder buffer · Exception · Branch misprediction · Issue queue · Wrap bit

1 Introduction

Reservation stations are a buffering unit on the border between the in-order front end and the out-of-order back end of microprocessors. They were invented as waiting stations where instructions wait for source operands to become ready or designated execution units to become available [1]. Instructions in the reservation stations arrive through an in-order multi-stage pipeline. In order to keep the pipeline full, microprocessors need to speculate and to fetch instructions that are speculated to represent the program order. In the case of miss-speculation, speculatively inserted instructions need to be flushed from the microprocessor. Then, instructions from the correct program order need to be fetched.

Figure 1 shows a microarchitecture of a superscalar core. The core comprises an in-order front end and out-of-order back end. The in-order front end may include a fetch unit, a decode unit, a renaming unit, and a dispatch unit [2]. The back end of the core may include one or more execution slices, one or more physical register files, and a retirement unit [3]. The core may support multiple instruction issue, in-order and out-of-order execution, and simultaneous multithreading, where instructions from up to two threads may simultaneously be processed.

© Springer Nature Switzerland AG 2020
V. Dimitrova and I. Dimitrovski (Eds.): ICT Innovations 2020, CCIS 1316, pp. 127–137, 2020.
https://doi.org/10.1007/978-3-030-62098-1_11

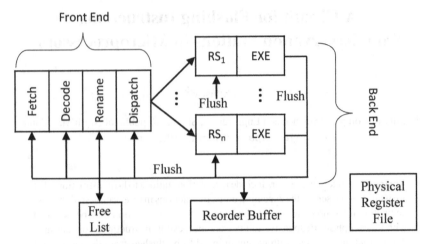

Fig. 1. Microarchitecture of superscalar core

The fetch unit is configured to fetch instructions from a memory or L1 cache. Instructions may be from any instruction set architecture, e.g. PowerPC™, ARM™, SPARC™, x86™, etc. The fetch unit includes a branch predictor to predict the taken branch. Responsive to a branch prediction, the fetch unit is fetching instructions from the predicted branch. The decode unit is configured to receive instructions from the fetch unit and to output decoded instructions or micro-operations, which are later executed in the execution units. The free list is a list of physical registers that are allocated to instructions with destination operands. For each instruction with destination operand, the free list allocates a physical register; thus, performing renaming of the destination operand to physical register. Allocated physical register is stored in a mapping table and used to rename source operands of younger data-dependent instructions. The renaming unit is configured to rename (map) source operands of instructions consumers of a result to the physical register allocated to the instruction producer of the result. A source operand of an instruction is renamed to the physical register most recently allocated to another instruction with destination operand equal to the source operand. Renamed instructions are provided to the dispatch unit. The dispatch unit sends the renamed instructions to the reservation stations RS_1 to RS_n.

The core includes one or more reservation stations RS_1 to RS_n. Each reservation station is coupled to an execution unit (EXE). Each execution unit includes any number and type of execution units, e.g. an integer unit, a floating-point unit, a load/store unit, a branch unit, etc. Execution units execute instructions. Instructions may be executed in-order or out-of-order. Out-of-order executed instructions are waiting for in-order retirement in the reorder buffer.

The back end of the microprocessor includes a reorder buffer. The reorder buffer maintains in-order retirement of the instructions executed out of program order. For each instruction, the reorder buffer allocates an entry and assigns the index of the allocated entry to the instruction. Reorder buffer entries are organized as a circular buffer with a tail pointer and a head pointer. Instructions enter at the tail pointer and exit at the head pointer. Thus, instructions can be executed out of program order but retired in program order. The core includes a large number of physical registers organized in one or more physical register files. Physical registers of the core store speculative results and architecturally visible results. The core may employ stand-alone addressing scheme for the physical registers, or physical registers may use the addressing scheme of the reorder buffer.

When an instruction is executed with an exception, the core handles the exception event by flushing the miss-speculated instructions and restoring the architectural state to the state prior to the exception. Physical registers allocated to miss-speculated instructions are returned to the free list. These physical registers are identified with a walk on the reorder buffer from the tail pointer to the reorder buffer entry of the instruction executed with the exception. The content of the mapping table is restored to the state prior to the exception. Mapping table restoration may be done with checkpoints [4] or with the walk on the reorder buffer. The tail pointer of the reorder buffer is moved to the instruction executed with exception. However, the pipeline remains stalled until reservation stations flush miss-speculated instructions. Reservation stations may use branch masks to flush younger instructions or may postpone the flush until all instructions older than the instruction executed with exception are retired.

In this paper, we propose a circuit for immediately flushing instructions from reservation stations. Responsive to exception event, the proposed circuit is configured to selectively flush all miss-speculated instructions. The circuit is configured to leave older instructions to be executed. Immediate flush of instructions may improve the performance of the processor for at least 2% [5]. Considering that 49% of all fetched instructions are flushed [6], it is important for reservation stations to be flushed as soon as possible. The remainder of this paper is organized as follows: the principle of operation of the reorder buffer is given in Sect. 2, a conventional flush logic, i.e. branch masks, is proposed in Sect. 3, the proposed circuit for flushing instructions from reservation stations is explained in detail in Sect. 4, experimental results from the comparison between the conventional flush logic with branch masks and the proposed circuit for flushing instructions is provided in Sect. 5, related work is discussed in Sect. 6. Section 7 concludes the paper.

2 The Reorder Buffer

Figures 2A and 2B demonstrate the principle of operation of the reorder buffer. The reorder buffer is a circular buffer with N + 1 entries indexed with 0 to N, where entry 0 is considered successor of the entry N. The reorder buffer is configured to maintain an entry for each in-flight instruction in the core. Two adjacent reorder buffer entries are allocated to two adjacent in program order instructions. Thus, instructions in the reorder buffer are considered to be in program order.

The reorder buffer includes a tail pointer to point to the entry that is next for allocation. For an instruction entering the core, the reorder buffer allocates this entry and increments the tail pointer. The reorder buffer includes a head pointer (not shown) to point to the oldest in-flight instruction. An instruction may be retired only if the head pointer points to the reorder buffer entry allocated to the instruction.

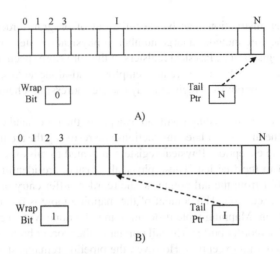

Fig. 2. Principle of operation of the reorder buffer

The reorder buffer is configured to maintain a wrap bit with the tail pointer. The wrap bit is toggled each time the tail pointer wraps-around from the entry N to the entry 0. Figure 2A shows the wrap bit holding logical 0. After the entry N is allocated, the wrap bit is toggled (see Fig. 2B). The wrap bit maintains logical 1 for the duration of the entire allocation cycle from 0 to N. The reorder buffer is configured to maintain separate wrap bit with the head pointer.

One application of the wrap bits is to determine whether the reorder buffer is full or empty [7]. If the tail pointer and the head pointer point to the same entry and the corresponding wrap bits are equal, then the reorder buffer is empty. If the tail pointer and the head pointer point to the same entry and the associated wrap bits are not equal, the reorder buffer is full.

Another application of the wrap bits is to determine relative age between two instructions. If wrap bits of two instructions are equal, then the instruction with smaller reorder buffer index is older. If wrap bits of two instructions are different, then the instruction with larger reorder buffer index is older. For example, an instruction with reorder buffer index EX and wrap bit WX is executed with exception. The reservation stations are configured to compare reorder buffer indexes and wrap bits of the hosted instructions with the index EX and the wrap bit WX in order to determine relative age between each of the hosted instructions and the instruction executed with the exception. Instructions younger than the instruction executed with exception are instantaneously flushed from the reservation stations, while older instructions are left to be executed.

Instructions that may initiate flush

	0	1	2	3	4
0	1	0	0	0	1
1	1	1	0	0	0
2	1	1	0	1	1
3	1	1	0	0	1
4	0	0	0	0	0
5	1	1	0	1	1
6	1	1	0	0	1

Reservation Station Entries

Fig. 3. Branch masks

3 Branch Masks

A reservation station with M entries may host up to M instructions. A flush logic may be coupled to the reservation station and configured to flush instructions younger than an instruction executed with an exception. The flush logic may be based on the branch masks of size M × K to maintain the relative age among instructions in the reservation stations and instructions that may initiate flush operation [8]. Figure 3 shows an example of 6 × 5 branch masks for a reservation station with M = 6 entries. Each row in the branch masks corresponds to an instruction in reservation stations. Each column in the branch masks corresponds to an instruction that may initiate flush operation if executed with exception. Instructions that may initiate flush operation are denoted with 0 to K − 1.

Branch masks may be considered a matrix of one-bit memory cells, where a memory cell B[i][j] holds the relative age between an instruction in an RS entry i and an instruction denoted with j that may initiate flush. For example, B[3][2] = 0 denotes that instruction in the RS entry 3 is older than the instruction denoted with 2 that may initiate flush. Following the column 1, we can conclude that the instruction denoted with 1 is older than the instructions in the RS entries 1, 2, 3, 5, and 6. If the instruction 1 is executed with exception, the flush logic will flush the instructions from the RS entries 1, 2, 3, 5, and 6, while leaving instructions in the RS entries 4 and 0.

In this paper, we consider branch masks to be of size MxM. Thus, any instruction in the reservation stations may initiate flush operation. This makes branch masks comparable with the proposed circuit for flushing instructions.

When an instruction is stored in the reservation station, a row-vector and a column-vector must be assigned to the instruction. The row- vector consists of a string of logical 1 s. The column-vector consists of a string of logical 0 s. Logical 1 s are written in the assigned row of the branch mask. Logical 0 s are written in the assigned column. This indicates that the stored instruction is the youngest instruction in the reservation stations. When two or more vectors enter the reservation stations, computing the row-vectors and column-vectors is more complex.

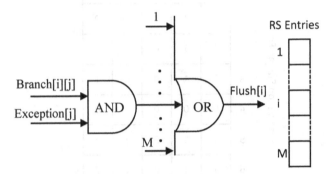

Fig. 4. Flush logic with Branch Masks.

Figure 4 shows the flush logic based on branch masks coupled to the reservation stations. The flush logic consists of M^2 AND gates and M OR gates with fan-in of $M - 1$. The M OR gates assert flush signals Flush[1] to Flush[M]. Asserted flush signal Flush[i] may flush the instruction hosted in the reservation station entry i. The flush logic is coupled to receive M signal lines Exception[1] to Exception [M]. Asserted Exception[j] signal indicates that the instruction in the reservation station entry j was executed with exception. Each AND gate is coupled to a memory cell B[i][j] from the branch masks and to an Exception[j] signal. Thus, if the instruction in the reservation station entry j was executed with exception and if B[i][j] = 1, the AND gate may output logical 1, which will trigger the OR gate to assert the flush signal Flush[i], which will flush the instruction in the reservation station entry i. Branch mask of size MxM implies that the hardware complexity of the flush logic is quadratic, i.e. proportional to M^2.

4   The Proposed Circuit for Flushing Instructions

Figure 5 shows the proposed circuit for flushing instructions coupled to a reservation station with M entries. In response to exception event, the proposed circuit for flushing instructions is configured to immediately flush instructions from the reservation station. Reservation station entries host instructions that consists of wrap bit W and index I. The wrap bit W and the index I are assigned by the reorder buffer. In a multithreaded core, each thread my include a thread-specific reorder buffer, which may assign indexes and wrap bits to instructions from one thread.

Given two instructions from one thread, instruction that precedes in program order is considered older. We will assume that the reorder buffer indexes are assigned in increasing fashion. If wrap bits of two instructions are equal, then the instruction with smaller index is older. If wrap bits of the two instructions are different, then the instruction with larger index is older. Responsive to exception event, e.g. branch misprediction, cache miss etc., a wrap bit WX and an index EX of an instruction executed with exception

RS Entries

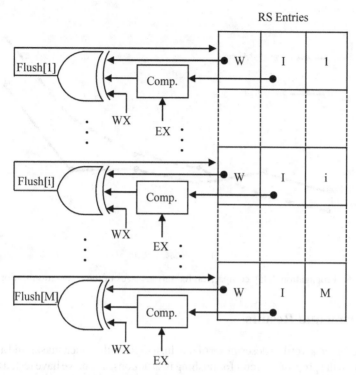

Fig. 5. Proposed circuit for flushing instructions from the reservation stations.

may be provided to the proposed circuit. The proposed circuit may be configured to compare indexes I of the instructions in the RS entries with the index EX, and to flush each instruction with wrap bit W equal to WX and index I larger than EX, and each instruction with wrap bit W not equal to WX and index I smaller than EX.

The proposed circuit is configured to output M flush signals Flush[1] to Flush[M]. Asserted Flush[i] signal may flush the instruction in the reservation station entry i. The proposed circuit consists of comparators and XOR gates. Comparators are configured to compare indexes I with the index EX. Each comparator is configured to output logical 1 if EX < I, or 0 otherwise. XOR gates are coupled to receive the wrap bit WX, wrap bits W from the reservation station entries, and output from the comparators. XOR gates are configured to output flush signals Flush[1] to Flush[M] which may initiate flush operation on the coupled entries. Asserted flush signal indicates that the instruction executed with exception is older than the instruction hosted in the corresponding reservation station entry. In a single-threaded core, asserted flush signal may flush the instruction hosted in the entry. In a multi-threaded core, additional test is performed to determine if the instruction in the reservation station entry and the instruction executed with exception are in the same thread.

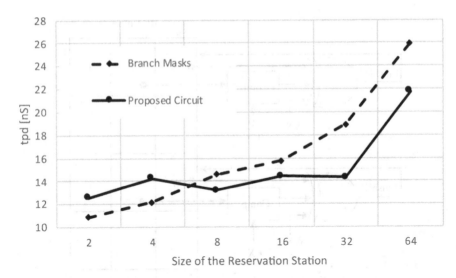

Fig. 6. Propagation delay comparison for various sizes of the reservation station.

5 Experimental Results

We have designed a Verilog description of the flush logic with branch masks and a Verilog description of the proposed circuit for flushing instructions. Then, we have tested both the flush logic with branch masks and the proposed circuit for flushing instructions using the Quartus software targeting the Cyclone FPGA family [9]. We have observed the propagation delay and the total number of elements for various sizes of the reservation station.

Figure 6 shows propagation delays of the flush logic with branch masks and of the proposed circuit for flushing instructions with respect to the number of entries in the reservation station. We use 9-bit comparators to compare reorder buffer indexes for the proposed circuit for flushing instructions. Reorder buffers in modern microprocessors are smaller than 512 entries, therefore 9 bits are enough to represent the address space of the reorder buffer. It can be observed that the proposed circuit shows better propagation delay for reservation stations with 8 or more entries. The flush logic with branch masks shows smaller delay for reservation stations with less than 8 entries. This is expectable since the proposed circuit with 9-bit magnitude comparators is more complex than the flush logic with branch masks for reservation stations with less than 8 entries.

Figure 7 compares the total number of logic elements synthetized for the proposed circuit for flushing instructions and the total number of logic elements synthetized for the flush logic with branch masks with respect to the number of entries of the reservation station. The total number of elements refers to the total number of combinatorial elements plus the number of registers. From Fig. 7 it can be observed that the total number of elements in the flush logic with branch masks grows quadratically with respect to the number of entries in the reservation station. This is expected result due to the quadratic hardware complexity of the branch masks. We can observe that the total number of elements in the proposed circuit for flushing instructions grows linearly with respect to the number of entries in the reservation station. Thus, for reservation stations with more than 8 entries the advantage of the proposed circuit is obvious.

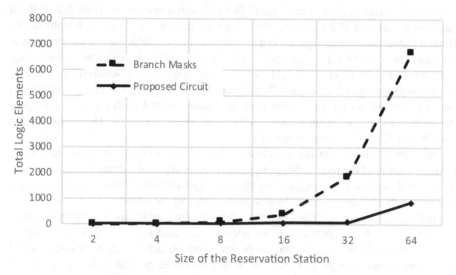

Fig. 7. Number of logic elements for various sizes of the reservation station.

6 Related Work

Reservation stations can be implemented as shifting FIFO-like structures, where instructions enter at one end of a queue, while the search for the oldest ready instruction starts at the other end. One such an example is the Pentium 4 processor [10]. The Alpha 21264 microprocessors are another such an example where reservation stations are implemented with collapsing buffers [11]. The position of an instruction with respect to the exiting end of the queue is indicator of the relative age of the instruction. This approach simplifies flushing due to misprediction. However, a constant shift of instructions is needed in order to make space for new instructions. Constantly shifting instructions is expensive operation in terms of gates, die area, wiring, and power consumption.

Due to the high complexity of the collapsing buffers, modern reservation stations are implemented as non-shifting structures. In non-shifting reservation stations instructions are kept in one entry until their issuance to the execution units. Flushing instructions from the non-shifting reservation stations may be implemented by assigning branch masks to the hosted instructions. Each bit in the branch mask indicates dependence with prior branch instruction that may initiate flush. The MIPS R10000 superscalar microprocessor uses four-bit branch masks [8]. Hence, the MIPS R10K microprocessor can support at most four branch instructions in the reservation stations. A fifth instruction will cause a pipeline stall. Therefore, branch masks are complex design solution that does not scale well with the size of the branch mask. Keeping the branch mask as small as possible introduces stalls at runtime. Moreover, the number and the type of instructions (e.g.

branches) that may initiate flush operation is predetermined at the design stage of the processor.

POWER8 and older microprocessors use a variant of the branch masks, which are called flush masks [12]. A core of the POWER8 microprocessor supports 224 in-flight instructions divided in 28 groups, where each group consists of up to 8 instructions. Branch instructions are always the last instructions in the group. For each instruction in the reservation stations a 28-bit flush mask is assigned. Flush masks do not introduce stalls at runtime; however, they may leave numerous holes in the 224 slots for instructions due to the grouping of instructions. To alleviate this, POWER8 microprocessors support two branches per group, and partial flush of a group. Partial flush further complicates the design of the reservation stations, and the processor in general. Therefore, starting from POWER9 flush masks are not used [3].

Another way to flush instructions from non-shifting reservation stations is to wait for the head pointer of the reorder buffer to point to the instruction executed with exception. Then, to a flush operation is initiated on the reservation stations. We believe that currently this approach is favored among the designers of microprocessors.

In [5, 13], we have published another circuit for flushing instructions from the reservation stations. Although both circuits perform the same function, they are based on different concepts. We believe that the proposed circuit is better than the circuit in [5] since it uses fewer comparators per reservation station entry. This makes the proposed circuit faster and more energy efficient.

7   Conclusion

We have proposed a circuit designed to immediately flush instructions from reservation stations after an exception event. The circuit is designed so that any instruction of any type executed with exception may initiate flush on younger instructions. In contrast, the alternative flush logic with branch masks is usually designed such that only at-design-stage predetermined type and number of instructions may initiate flush operation. The proposed circuit exhibits acceptable propagation delay and hardware complexity. Hence, we believe it is suitable to be implemented in modern microprocessors.

Moreover, we believe that the proposed circuitry may harden processors against side-channel attacks [14]. Side channel attacks exploit speculative fetching and out-of-order execution to execute malicious instructions. For example, an instruction may be executed with an exception. A malicious instruction capable of leaking data through side-channel may be waiting in the issue queue. The proposed solution will immediately kill the malicious instruction.

References

1. Tomasulo, R.M.: An efficient algorithm for exploiting multiple arithmetic units. IBM J. Res. Dev. 11(1), 25–33 (1967)
2. Smith, J.E., Sohi, G.S.: The microarchitecture of superscalar processors. Proc. IEEE 83(12), 1609–1624 (1995)

3. Sadasivam, S.K., Thompto, B.W., Kalla, R., Starke, W.J.: IBM Power9 processor architecture. IEEE Micro **37**(2), 40–51 (2017)
4. Akkary, H., Rajwar, R., Srinivasan, S.T.: Checkpoint processing and recovery: towards scalable large instruction window processors. In: Proceedings of the 36th Annual IEEE/ACM International Symposium on Microarchitecture, MICRO-36, San Diego, CA, pp. 423–434. IEEE (2003)
5. Spasov, D.: An improvement in the convergence of superscalar processors. In: Proceedings of the 43rd International Convention on Information, Communication, and Electronic Technology, Opatija, Croatia (2020)
6. Naresh, V.R.K., Sheikh, R., Perais, A., Cain, H. W.: SPF: selective pipeline flush. In: Proceedings of the IEEE 36th International Conference on Computer Design (ICCD), Orlando, FL, pp. 152–155. IEEE (2018)
7. Papworth, D.B., et. al.: Entry allocation in a circular buffer using wrap bits indicating whether a queue of the circular buffer has been traversed. US Patent #5584037, USPTO (1995)
8. Yeager, K.C.: The MIPS R10000 superscalar microprocessor. IEEE Micro **16**(2), 28–41 (1996)
9. Intel Download Center for FPGAs. https://www.intel.com/content/www/us/en/programmable/downloads/download-center.html. Accessed Apr 2020
10. Hinton, G., et. al.: The microarchitecture of the Pentium® 4 processor. Intel Technol. J. (2001)
11. Kessler, R.: The Alpha 21264 microprocessor. IEEE Micro **19**(2), 24–36 (1999)
12. Sinharoy, B., et al.: IBM POWER8 processor core microarchitecture. IBM J. Res. Dev. **59**(1), 1–21 (2015)
13. Spasov, D.: Method and apparatus for flushing instructions from reservation stations. US Patent #10095525, USPTO (2018)
14. Horn, J.: Reading privileged memory with a side-channel. googleprojectzero.blogspot.com. Accessed Jan 2018

Parallel Programming Strategies for Computing Walsh Spectra of Boolean Functions

Dushan Bikov[1]([envelope])[iD] and Maria Pashinska[2][iD]

[1] Faculty of Computer Science, Goce Delchev University, Shtip, Macedonia
dusan.bikov@ugd.edu.mk
[2] Institute of Mathematics and Informatics, Bulgarian Academy of Sciences,
Veliko Tarnovo, Bulgaria
mariqpashinska@math.bas.bg

Abstract. The utilization of all computational resources is significant to achieve efficient computation. In order to exploit the available computation resources, we combine two parallel programming models such as MPI and CUDA. Combining of these two programming models ensures usage of the whole computation resources available in one computer system (CPU and GPU). In this paper, we present a way to use the available parallel processing resources to their full potential utilizing different strategies and techniques regarding data transfer. We perform experimental computation of well know algorithm for computing Walsh spectra of Boolean functions by combining these two parallel programming models. Experiments are performed on two different class of parallel processing capability hardware. Randomly generated Boolean functions of fourteen, sixteen, eighteen and twenty variables represent the used data set for experiments evaluations. Performed experiments show how the growth of the data size results in gaining more parallelization and therefore accelerate the execution.

Keywords: MPI · CUDA · Boolean function · Walsh spectra · Parallel algorithm

1 Introduction

In general software has been written for sequential computation. Carried by the latest achievements in computing technologies we could use computation resources more efficiently. The widely used computing systems combine two main computational resources. It's well known that CPU (Central Processing Unit) is

The research of the first author was supported by Bulgarian Science Fund under Contract DN-02-2/13.12.2016.
The research of the second author was supported, in part, by the Bulgarian Ministry of Education and Science by Grant No. DO1-221/03.12.2018 for NCHDC, a part of the Bulgarian National Roadmap on RIs.

V. Dimitrova and I. Dimitrovski (Eds.): ICT Innovations 2020, CCIS 1316, pp. 138–152, 2020.
https://doi.org/10.1007/978-3-030-62098-1_12

the heart of the computation system but recently modern GPU (Graphics Processing Unit) takes significant part which is due to its massive parallel processing capability that it owns.

The main goal of this paper is to show how we can utilize all computation resources by using two different class of parallel processing hardware. Modern CPU is power efficient processor with multiple cores. Recently GPU have highly parallel structure which allows efficient parallel processing of large blocks of data. Nowadays, the computer integrates and combines the good properties of CPU and GPU in one system. Multi-core processors have brought parallel computing to wide use general-purpose personal computers, embedded system, workstations, game consoles, smart phones etc.

Led by curiosity and our scientific interests in Cryptography and Coding Theory the implementation of the well-known problem will present bigger picture for the behavior and using of modern technologies. For the purpose of exploiting all computation resources we use the well-known problem for computing Walsh Spectrum of a Boolean function. Walsh Spectrum can be computed by Walsh (Hadamard, Walsh-Hadamard, Walsh-Fourier) transformation which has a wide range of applications. It is used in cryptography, signal and image processing, image rendering, data compression algorithms, etc. There are many algorithms for computing of Walsh Spectrum but for our purpose we use the algorithm described in paper [1].

The idea is to use different programming strategies by combining MPI (Message Passing Interface) and CUDA (Compute Unified Device Architecture) programming models. MPI is a standard API (Application Programming Interface) and well-known programming model for communicating data via messages between distributed processes or in other words for Distributed Memory Computing. As such, MPI is fully compatible with CUDA, which is parallel computing platform and API model created by Nvidia [11]. CUDA programming model allows software developers to use a CUDA-enabled GPU for general purpose processing. This platform is a software layer and gives direct access to the GPU to run programs on them. We can gain access to the GPU resources by using CUDA. MPI can be used for distributing the tasks and every MPI thread would have CUDA instance.

There are other approaches and techniques for combine use of CPU and GPU resources. These approaches are implemented through specific programming language, API or software framework [8] for writing programs that execute across heterogeneous platforms consisting CPU, GPU, DSP (Digital Signal Processors) or hardware accelerators. Some of the solutions are intended for cluster computing, others have adjustable parallelization capability. Critical for choosing Nvidia GPU was the significant effort that they invest in technological improvements. Available computer systems (Table 2) predetermined the use of MPI and CUDA programming models in order to achieve ours goals. There are various technique for concurrent kernel execution for reclaiming of lost performance or wasted resources. In CUDA this is supported by the stream interface [16]. Streams act as independent queues through which can executed different kernels simultaneously. They improve overall concurrency but also exhibited signs of false serialization where independent streams might not overlap because of queuing order, memory copy serializations and interleaving. This problem is overcome by Hyper-Q/MPS (Multi-Process Service) technology which allows independent, hardware-managed work queues [9]. This technology allows CUDA kernels to be processed concurrently on the same GPU on the other hand our goal is to improve memory transfer without use of streams and to show the impact on concurrent kernel execution. In some way our idea is close to the new MIG (Multi-Instance GPU) [14] feature of the Nvidia A100 GPU [10].

The paper is organized as follows. The main definitions connected with Boolean functions and Walsh spectrum are given in Sect. 2. In Sect. 3 we describe general principles of MPI and CUDA programming model and possibility for combining both of them in a single program. Section 4 is devoted to the parallel models strategies and algorithms for computing Walsh Spectrum. Results from experimental evaluation are presented in Sect. 5. Few conclusion sentences is given in the end.

2   Boolean Functions and Walsh Spectrum

Boolean functions are basic objects in Discrete Mathematics, and they have important role in Cryptography, Signal and Image Processing, Coding Theory, Data Compression Algorithms, etc.

A Boolean function f of n variables is a mapping from $\mathbb{F}_2^n$ into $\mathbb{F}_2$, where $\mathbb{F}_2 = \{0, 1\}$ is the field with two elements. Boolean function has many representations and two naturals are its $TT(f)$ (Truth Table) and $ANF(f)$ (Algebraic Normal Form) [3]. The $TT(f)$ is the 2^n-dimensional vector which has the function values of f as coordinates. The vectors in $\mathbb{F}_2^n$ can be considered as binary representations of the integers in the interval $[0, \ldots, 2^n - 1]$. This consideration is very useful for the description and explanation of some transformations of Boolean functions and related algorithms. Here efficient algorithms for calculating the Walsh spectrum are used. The basis is a matrix and vector multiplication. In this case the considered matrices have not only recursive structure, but this structure is quite specific and enables a very effective (*butterfly*) multiplication.

The function $(-1)^f = 1 - 2f$ whose values belong to the set $\{1, -1\}$ is associated with the Boolean function f and correspond vector that contains the functions values is called the Polarity Truth Table of the Boolean function f ($PTT(f)$).

Walsh transformation f^W of the Boolean function f is the integer valued function $f^W : \mathbb{F}_2^n \to \mathbb{Z}$ [3], defined by

$$f^W(a) = \sum_{x \in \mathbb{F}_2^n} (-1)^{f(x) \oplus f_a(x)} = \sum_{x \in \mathbb{F}_2^n} (-1)^{f_a(x)} (-1)^{f(x)},$$

where $a = (a_1, \ldots, a_n) \in \mathbb{F}_2^n$, $f_a(x) = a_1 x_1 \oplus a_2 x_2 \oplus \ldots \oplus a_n x_n$ and $f(x) \oplus < a, x >= f(x_1, x_2, \ldots, x_n) \oplus a_1 x_1 \oplus a_2 x_2 \oplus \ldots \oplus a_n x_n = f(x) \oplus f_a(x)$.

The values of f^W are called $Walsh\ coefficients$. Lets consider the vectors of $\mathbb{F}_2^n$ ordered lexicographically. Then the Walsh coefficients $f^W(a)$ can be ordered and considered as coordinates of a vector. This vector is called Walsh spectrum of the Boolean function and is denoted by $[W_f]$ (considered as a column). Walsh spectrum $[W_f]$ can be obtain as multiplication of $PTT(f)$ and Hadamard Matrices (Sylvester type) H_n from order n [7].

In our case for calculating Walsh spectra of Boolean function f we use Fast Walsh Transform sequential base on Algorithm 2, given by a butterfly diagram (Diagram 2 [2]) and the corresponding algorithm with complexity $O(n2^n)$ (Fig. 3 [5]). For the parallel computation a modified version of Algorithm 6 from [1] is used with added tricks and techniques explained in Sect. 4.

3   CPU and GPU Computing Model

One way to understand difference between CPU and GPU is to compare how they perform tasks. CPU have few cores optimized for sequential serial code execution while the GPU have architecture that consists of thousands very efficient small cores designed for processing thousands of tasks simultaneously.

Generally, MPI is a standard API for exchange data between processes via messages. There are several open source and commercial implementations of CUDA-aware MPI such as MPICH, OpenMPI, MVAPICH, IBM Platform MPI, Cray MPI etc. with bindings for C/C++, Fortran, Python etc. MPI is fully compatible with CUDA, designed for parallel computing on single computing system. This distributed memory programming model is commonly used in HPC (High Performance Computing) to build applications that can scale to multi-node computer clusters. There are many reasons for combining these two parallel programming models. Our interest is to accelerate an existing sequential or MPI application with GPU and using CUDA-aware MPI we can achieve this goal effectively.

MPI standard defines the syntax and semantics of a core of useful library routines for writing a wide range of portable message-passing programs in C/C++, Fortran, Python etc. Processes involved in an MPI program have individual computing represented by an abstract connection of the processing element to its own address space in which it is run. Each processing element or MPI process is called rank. For the duration of the program execution there are fixed number of ranks and all of them execute the same program also know as SPMD (Single Program Multiple Data) [15]. Each rank runs on different core and has private memory. Data can be copied or moved into the private memory of the rank. There are two MPI communications schemes, point-to-point communication (communication between two processes) and collective communication (involves participation of all processes). In general all ranks either communicate or compute and all ranks perform the same activity at the same time. However, ranks workloads aren't well balanced. It is fundamentally important to understand passing messages. They are like email, we define destination and message body (can be empty). Communication is two sided and requires explicit participation of sender and receiver. The messages provide memory-to-memory across address spaces and synchronization of 2-sided handshake (communication is completed, and data has arrived). If the rank sends multiple messages to the same destination, order of receiving will be as the order of sending. But if different ranks send messages to same destination the order of receiving can't be determined. Library called MPI is used for writing MPI programs. The first release is MPI-1 in 1994 and it includes 125 routines. MPI-3 consists of more than 430 routines. Most MPI programs need at least these 6 routines: start, end, query MPI execution state, point-to-point message passing. In addition, MPI library has tool for launching an MPI program (mpirun) and MPI daemon for moving data over the network.

CUDA is a powerful parallel computing platform and API created by NVIDIA for general purpose computing using GPU [11]. This is a computing platform that allows developers to interact directly with the GPU and speed up computing applications by harnessing parallel computing resources. GPU accelerated application contains two parts, the sequential (control) part that runs optimized single CPU thread and the GPU executed part that runs thousands of cores in parallel. The application can be written in programming languages such C/C++, Fortran, Python and MATLAB and they express parallelism through extensions by few keywords. Depending on the GPU architecture CUDA cores are organized in SM (Streaming Multiprocessor) and the memory has hierarchical ordering that consists of registers, cache for constants, texture cache (read-only arrays), shared memory (L1 cache) and global memory. Every SM has a few SP (Streaming Processor), and a few SFU (Special Function Unit) used for transcendental functions as sine, cosine etc. Common name for SP is CUDA core. Each SP includes several ALU (Arithmetic Logic Unit) and FPU (Floating Point Unit). Each SM executes instruction in a SIMT (Single-Instruction Multiple-Threads) [6] mode which is similar to the SIMD (Single Instruction Multiple Data) architecture, and the communication between SM is performed through global memory.

On the top level of the CUDA application there is a master process that is run on the CPU which is responsible for initialization of the GPU card, allocation of the memory and moving between main memory and GPU memory, launching of the kernels on the GPU responsible for performing the computations, fetching back the memory once the computation is completed, deallocation of the memory and termination.

4   Parallel Programming Strategies

There are many possible tricks, techniques and strategies that can be used in order to write efficient code. In this section we will explain what kind of tricks, techniques and strategies we have implemented in order to obtain more efficient code and to utilize more computation resources. Some of the improvements that are explained here will be supported with experimental evaluations and others will be commented in the result general experimental evaluations in the next section. The next discussion is in reference to CUDA programming model.

---

fwt_kernel$(TT_{int}, W_f, block\_size)$ Modify Algorithm 6 [1], row with conversion

---

Input: The integer array TT_{int} with $(2^n)/32$ *int* entries, and *block_size*
Output: The array W_f with 2^n output elements

...

$i_{int} \leftarrow tID/32;$ /*additional index for conversation*/
$i_{int}Rsh \leftarrow tID\%32;$ /*additional index for conversation*/
$value_{in} \leftarrow TT_{int}[i_{int}];$ /*local variable for intermediate input */
$value_{in} \leftarrow (value_{in} \gg i_{int}Rsh)\%2;$ /*conversion */
$value \leftarrow -1 \times (value_{in} - 1 + value_{in});$ /*polarisation*/

...

The biggest problem that occurs in GPU programming model is memory transfer from host to device and vice versa. Because of this the first question was what can be done in order to optimize the data transfer. The problem is related to Boolean functions and suitable way to transfer data from host to device is needed. One acceptable solution is to convert $TT(f)$ into dynamic array of unsigned integers and to transfer the data. In practice, there is a kind of binary to integer conversion (Table 1) before sending the data. Firstly, the TT is separated into 32 (or 64) long input vectors. Every element of this vector according the index is set to the appropriate bit position into the integer. The converted data occupy the same amount of memory regardless of the used data type (32 or 64 bits integers). Some experiments over data types, allocation and transfer of the memory are also performed. In the first case the experiments are executed with $32 - bits$ integers ($unsigned\ int$) and $64 - bits$ integers ($unsigned\ long\ long\ int$) and the conclusions are the following: transfer of the $32 - bits$ integers takes less time for the same amount of transferred data and it varies for the different amount of data but in best case is half. Pageable/pinned memory host allocation and data transfer experiments are preformed next. They were executed on Platform 1 (Table 2). Taking into account the appropriate conditions, such as amount of memory transfer, data type, etc. the results didn't show some significant time distinctions as expected. Because of these reasons we use default host memory allocation (pageable) in all other experiments.

It is important to noted that reverse conversion and polarization of the converted elements into the CUDA kernel (**fwt_kernel** modify algorithm) does not affect computational performance. As it is explained in the second section the result practically is 2^n long integer array.

Depending on the size of the Boolean function, different integer data types can be used for storing the results. Data type with smaller bit width will reduce time for data transfer from device to host. Required time for performing this action for transfer the data will be less. According to this data type *short int* can be used for Boolean function f when $n \leq 15$ and data type *int* for $n \geq 16$. The usage of *short integer* requires less time for data transfer by variation around 15% to 25% depending on the Boolean size in comparison to when we use only *int* for every case. This time benefit from the different data type we have in basic CUDA programming model. When combination of CUDA and MPI is used the benefit is almost unnoticeable because of the context switch overheads.

Some of the experimental tests were performed on Windows operation system (Windows 10 Pro), Microsoft Visual Studio 2013, same version of CUDA and close Nvidia drivers version on same hardware on both Platforms (Table 2). From the observation and the experimental evaluation it can be concluded that performance of data transfer is more efficient on Linux.

Table 1. $TT(f)$ to integer conversion

$TT(f)$ input vector (32-bits)	00110101100001000011010110000100
Unsigned int (32-bits)	3482526124
$TT(f)$ input vector (64-bits)	0011010110000100110010011110011 1111001001010010101010011101010001
Unsigned long long int (64-bits)	10008487453546193324

4.1 Parallel Models

In this subsection we will be briefly explain the used parallel strategy models with their advantages and disadvantages. The parallel strategy model here in general refers to the used techniques for reading the data which is strongly connected with the data transfer. Here we will explain the typical features of three models. The structure of our programs is simple and have several MPI processes that are sharing the same GPU. It is known that this structure introduces context switch overheads. The goal here is to see how the models behave under the same conditions and to find out which one is the most efficient. Figure 1 shows the general structure of processing flow together with execution model on all strategy separately.

First model Basic Parallel Strategy Model is shown on Fig. 1 marked by 1. As it is referred by the name we rely on this model and on its basis further development is made. This model has all the typical basic characteristic of a simple bound MPI-CUDA program. The main problem that CUDA have with data transfer is reflected here and is not possible to bypass or ignore. Here a simple solution for reading the input data is used. Because our input data is determined number of $TT(f)$, it can be split into separate equal input text files. The number of input files depends on the numbers of called MPI process. According to the MPI process ID (rank) for all of them it is assigned separate input data file. Every MPI process on a basic level reads one row which represent the Boolean function $TT(f)$, performs conversion (as is explained earlier), transfers the data to device memory and calls CUDA kernel which computes Walsh spectra. After the finish of computation of one Boolean function the result is transferred from device to host memory and the MPI process reads the next Boolean function $TT(f)$, performs the same steps for the new function. This procedure is the same for all process and is repeated until last row in the input file. All MPI processes repeat this procedure simultaneously. Even though the input files for all processes are separated, there is one hardware storing resource HDD (Hard Disk Drive) from where all process try to read simultaneously. The attempts of more processes simultaneously to gain access to the same hardware resource leads to resource contest, waiting, overheads and slows down the data transfer. With the increasing number of processes the negative effects are more pronounced and the benefit from parallel execution is gradually lost. The experimental evaluation of this technique is described in the next section.

The second model that will be described is Master-Workers Parallel Strategy Model shown on Fig. 1 marked by 2. This model has the well-known Master-Worker strategy. In general Master-Workers Strategy has one master process that schedules computational tasks on the workers. Here the master process reads the input data from file and sends pieces of that data for further computation. The input file at each row contains Boolean function $TT(f)$ and the number of rows gives the number of input Boolean functions. The master process on a basic level reads a Boolean function TT(f) and sends it to the next free worker process. After receiving the $TT(f)$ vector the worker process performs conversion (as is explained earlier), transfers data to the device memory and calls CUDA kernel where Walsh spectra is computed. Master process will continue reading until the last row of the input file. Even though there is one master process for reading input data from the storing hardware resource and for scheduling the significant improvement relate to data transfer is not achieved. With the growth of the Boolean function this observation becomes more obvious. On the other hand, conditionally speaking an extra process is used for reading the input data. The use of an extra process relieves access to the hardware storing resource. This strategy also leads to input query, task waiting time and not equal distribution of tasks between the workers. Experimental evaluation will be presented in the next section.

At last we will talk about One Input File Parallel Strategy Model shown on Fig. 1 marked by 3. One Input File Strategy in some way represents extension of Basic Strategy and the main difference here is the usage of one input data file and the approach for reading input data. The input file has the same structure each row contain Boolean function $TT(f)$ and the number of rows shows the number of input Boolean functions. All MPI processes on a basic level read different row from the same input file. To be able to perform the read and do not overlap the rows we use conditional statement as it is shown below:

$$if(NumRow\%nprocs == myrank);$$

Here $NumRow$ is the number of input Boolean function, $nprocs$ is the total number of running MPI processes and the $myrank$ is a MPI process ID. Here there is one input file that is open to all running MPI processes that try to read data at the same time. Once the row is read from the process the data is transferred to the device memory and CUDA kernel is called to compute the Walsh spectra. After the finish of computation of one Boolean function the result is transferred from device to host memory and the MPI process reads the next Boolean function $TT(f)$ according the parameter mentioned earlier and continues with its work. This proceeds for all process until they reach the end of the input file. Through experimental evaluation we will see the behavior of this model.

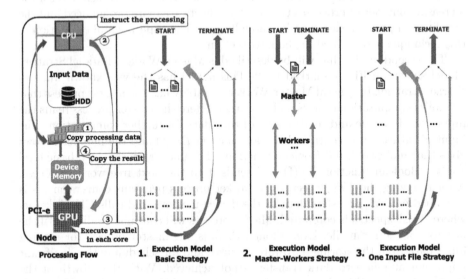

Fig. 1. Example of processing flow and execution strategy model

It is completely possible to use the explained parallel strategy models for building standalone MPI programs. The essential part of those strategies is the way input data is read. Practically everything is the same except that input data is computed from the MPI processes themselves. In the first and the third strategy the process that reads the data does the computation. In the second strategy the process worker does computation. In all of the strategies at the process level it is used the same algorithm for computation of Walsh spectra [2].

5   Experimental Evaluation

In this section we present our experimental results. The platforms that are used for the experimental evaluation are described in Table 2. Platform 1, a graphic card NVIDIA GeForce 150MX [12], has 384 cores running at 1.53 GHz and 48 (GB/s) memory bandwidth. Platform 2, a graphic card NVIDIA TITAN X [13], has 3584 cores running at 1.41 GHz and 480 (GB/s) memory bandwidth.

For building on single MPI + CUDA program CUDA-aware MPI implementations are used, in this case OpenMPI, Cuda-aware MPI implementation on Ubuntu OS. MPI can handle more than one GPU but in this case only one GPU is utilized which is shared between the several MPI process.

Table 2. Description of the test platforms tables

Environment	Platform 1	Platform 2
CPU	Intel i7-8565U, 1.80 GHz	Intel Xeon E5-2640, 2.50 GHz
Memory	8 GB DDR4 2400 MHz	48 GB DDR3 1333 MHz
GPU	GeForce MX150	Nvidia TITAN X (Pascal)
OS	Ubuntu 16.04 LTS 64-bit	Ubuntu 18.04.3 LTS 64-bit
Compiler	gcc 5.4.0	gcc 7.4.0
CUDA compilation tools	10.1	9.1
GPU Driver	V416.56	V430.50
MPI	(Open MPI) 1.10.2	(Open MPI) 3.3a2

There have been made different levels of experimentations where all explained parallel strategies were covered, in addition to MPI program, CUDA program [1] (**fwt_kernel** modify algorithm) and traditional CPU program [2]. For performing the experiments there was new version of CUDA program with added tricks and techniques mentioned in previous section. The MPI programs have used the same algorithm as in [2] but with the typical MPI features. Programs are build/executed in Release mode. Making object files, linking, building and executing of the programs were performed by using Ubuntu terminal and suitable commands. Our experimental evaluation mainly covers specific predefined input data set of 100 Boolean functions. To measure the execution time, the program is executed under same conditions and the time is obtained by using timers specially placed on suitable positions. However, in case of repetition of the experiment deviation of ±5% in measuring time of the execution may occur due to the hardware issues, operating system kernel, program execution, features of the programming model and the strategy itself. The input data set is contained in files that have determined structure which consists input binary arrays with 2^n entries ($n = 14, 16, 18, 20$).

As already mentioned the biggest problem that we deal with was memory transfer. We already pointed out used tricks and techniques which can be applied to alleviate the impact of the problem. First we will begin with experimental evaluation of data transfer on CUDA program and combined MPI + CUDA programs. The experimental data obtained by the upper mentioned timers is used to determine the time required for memory transfer ($cudaMemcpy$ [4]) between host and device.

Figure 2 shows the required time for memory transfer from host to device and vice versa on Platform 1 for given size of Boolean function (2^{14}). First part of the figure represent experimental evaluation when *short int* is used for transfer the data from device to host. The second part represents experimental evaluation when *int* is used for transfer the data in both directions. Transfer size in Bytes are show on the figure (see Fig. 2). As we can see in the Fig. 2

the experiment included all strategies and CUDA general version. Also there was separate experiments for all of the strategies with 2 and 4 MPI + CUDA processes. For performing these experiments the same input data is used. It consists of one hundred Boolean functions with same size and depending on the case there was distributed between the process. Time shown here is the total time for data transfer in both directions which is actually the average time from the performed experiment over the input data set. What can be noticed is that even though *short int* data type is used with half the size of *int* there are no significant performance improvements. It can point out that for same amount of data that is transferred there is minimal *cudaMemcpy* performance distinctions.

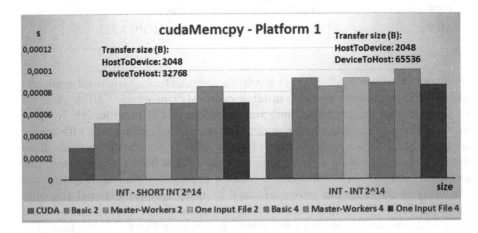

Fig. 2. cudaMemcpy Platform 1, *short int* vs. *int*

On Fig. 3 is shown more completely experimental evaluation of *cudaMemcpy* for Platform 1 where experiment for different sizes of Boolean functions $n = 14, 16, 18, 20$ are included. Here is shown only the use of *int* data type. Transfer size in Bytes are shown on the same Fig. 3. There was experiment with all strategies with 2 and 4 MPI + CUDA processes and CUDA general version. The time shown here is the full time for data transfer in both directions. The experiments are performed under same specific conditions depending on the case. In addition to previous conclusions, here it can be added that *cudaMemcpy* it is not stable for all of the strategies and strongly depend from amount of data transfer and of course the amount of computation that is perform in GPU. All of the strategies have typical features some of them have been discussed in the previous sections which are manifested differently under certain conditions. The mentioned dependency is strongly connected with available hardware resources and they are in strong correlation. If we analyze the last case ($n = 20$) for Master-Worker *cudaMemcpy* have best performance comparing to the other two strategies. Specific structure factors of Master-workers strategy under this conditions of execution with this amount of data transfer and computation give best performance comparing to others strategies.

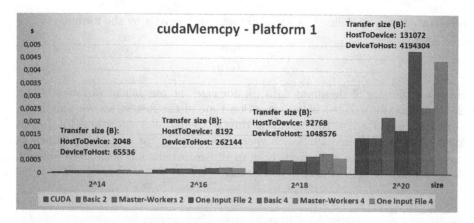

Fig. 3. cudaMemcpy Platform 1

The behavior of Platform 2 is similar to Platform 1 related to data transfer when *int* data type with *short int* is used. On Fig. 4 is shown completely experimental evaluation of *cudaMemcpy* for Platform 2 and the experiments are performed for different size of Boolean functions $n = 14, 16, 18, 20$. As can be seen we have a similar situation with Platform 1. Here we have slight increase/decrease of average times for data transfers which is due to slightly older technology of Platform 2. This distinctions also depends on the amount of data that is transferred, hardware resources capabilities, memory specifications etc. Behavior of the strategies in general is more stable without significant differences except case when $n = 20$. Everything that was said about this case for Platform 1 is also valid here.

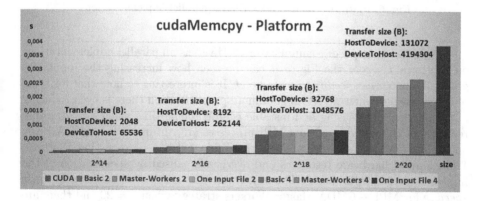

Fig. 4. cudaMemcpy Platform 2

By designing parallel program that will run faster than sequential program it achieves acceleration of parallel regarding the sequential implementation. This acceleration can be defined as *speedup* that can be given by the formula:

$$S_P = \frac{T_0(n)}{T_p(n)}.$$

Here n is the size of the input data (in our case the number of variables of the Boolean function), $T_0(n)$ is the execution time of the fastest known sequential algorithm, and $T_p(n)$ is the execution time of the parallel algorithm [15]. In the following it is presented Platforms *speedup* summary. For $n = 14$ experimental examples for transfer data from host to device *int* data type is used and for device to host *short int* data type is used.

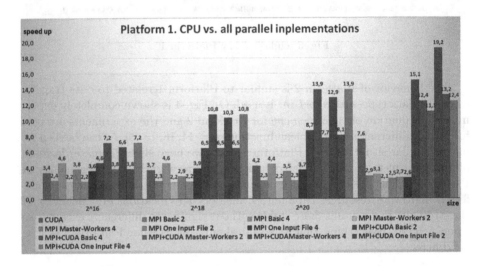

Fig. 5. Platform 1 summary, CPU vs. parallel implementations

Figure 5 shows *speedup* summary of CPU versus all parallel implementations on Platform 1. From the Fig. 5 it can be seen how increasing the input size gradually increases *speedup* to some point. It is interesting to note that for some of the strategies when we try to gain more parallelization there is even reducing of the *speedup*. Increasing and instable *cudaMemcpy* time when $n = 20$ imply reducing of the *speedup* when 4 MPI processes are run in MPI + CUDA Basic and Input One File strategies. From the Table 2 it can be seen that Platform 1 has limited hardware resources and overheads caused by specification of the implemented strategies occur. The best result that is achieved here is ×19 times *speedup* for MPI + CUDA Master-Workers strategy when $n = 20$ and there are 4 MPI processes.

Figure 6 shows *speedup* summary of CPU versus all parallel implementations on Platform 2. The situation here is significantly different comparing with Platform 1. Here no negative *speedup* appears and we gain more parallelization. This is due to hardware resources at Platform 2 disposal. All strategies have close performance with same conditions for most of the cases. The best result that is achieved here is ×54 times *speedup* for MPI + CUDA Master-Workers strategy when $n = 20$ and there are 4 MPI processes.

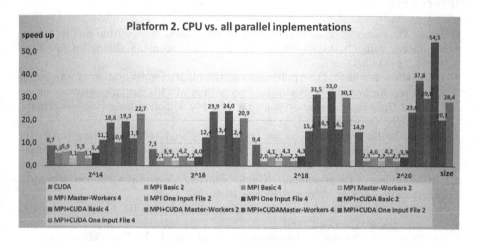

Fig. 6. Platform 2 summary, CPU vs. parallel implementations

As we expected Platform 2 gives better results compering to Platform 1. This is due to the fact that Platform 2 despite having slightly older technology, it has much better hardware and processing capabilities (see Table 2).

Briefly we will mention without further analysis that there were performed experiments under significantly small input data set that contain few Boolean function. From these experiments can conclude that the strategy models are more stable and it gives better execution times. This instability is mainly expressed in data transfer which from the other hand significantly affects performances.

6 Conclusions

In this paper, we compared the presented parallel strategies with different and combined programming models. No matter of the introduced MPI + CUDA strategies structure the problem with context switch overhead remains.

It is obvious that significant time is spent for data transfer. Of particular interest is the use of any kind of tricks, technique and strategies for optimising the data transfer. Here it is presented the worst case scenario when we have frequent transfer of memory from the host to device and vice versa. In general case for the CUDA programming and similar model it is recommended to have transfer of large amount of data for massive computations after that returning the results.

Growing of the size of input data and gaining more parallelization result in increasing of the *speedup* in favor of the combined parallel implementations.

Acknowledgments. We gratefully acknowledge the support of NVIDIA Corporation with the donation of the Titan X Pascal GPU used for this research.

References

1. Bikov, D., Bouyukliev, I.: Parallel fast Walsh transform algorithm and its implementation with CUDA on GPUs. Cybern. Inf. Technol. Cybern. Inf. Technol. **18**(5), 21–43 (2018)
2. Bouyukliev, I., Bikov, D.: Applications of the binary representation of integers in algorithms for boolean functions. In: Proceedings of 44th Spring Conference of the Union of Bulgarian Mathematicians, pp. 161–162. Union of Bulgarian Mathematicians, SOK Kamchia (2015)
3. Carlet, C.: Boolean functions for cryptography and error correcting codes. In: Boolean Models and Methods in Mathematics, Computer Science, and Engineering, vol. 2, pp. 257–397 (2010)
4. CUDA C++ Programming Guide. https://docs.nvidia.com/cuda/cuda-c-programming-guide/. Accessed 16 Aug 2020
5. Joux, A.: Algorithmic Cryptanalysis. Chapman & Hall/CRC Cryptography and Network Security Series (2009)
6. Lindholm, E., Nickolls, J., Oberman, S., Montrym, J.: NVIDIA Tesla: a unified graphics and computing architecture. IEEE Micro **28**(2), 39–55 (2008)
7. MacWilliams, F.J., Sloane, N.J.A.: The Theory of Error-Correcting Codes, vol. 16. Elsevier, Netherlands (1977)
8. Mittal, S., Vetter, J.S.: A survey of CPU-GPU heterogeneous computing techniques. ACM Comput. Surv. (CSUR) **47**(4), 1–35 (2015)
9. Multi-Process Service, GPU Deployment and Management Documentation. https://docs.nvidia.com/deploy/mps/. Accessed 16 Aug 2020
10. NVIDIA A100 Tensor Core GPU Architecture (whitepaper). https://www.nvidia.com/content/dam/en-zz/Solutions/Data-Center/nvidia-ampere-architecture-whitepaper.pdf. Accessed 16 Aug 2020
11. Nvidia CUDA Home Page. https://developer.nvidia.com/cuda-zone. Accessed 16 Aug 2020
12. NVIDIA GeForce 150MX Specification. https://www.geforce.com/hardware/notebook-gpus/geforce-mx150. Accessed 14 Apr 2020
13. NVIDIA GeForce TITAN X Specification. https://www.nvidia.com/en-us/geforce/products/10series/titan-x-pascal/. Accessed 16 Aug 2020
14. NVIDIA Multi-Instance GPU User Guide. https://docs.nvidia.com/datacenter/tesla/mig-user-guide/index.html. Accessed 16 Aug 2020
15. Quinn, Michael J.: Parallel Programming in C with MPI and OpenMP, 1st edn. McGraw-Hill Inc., New York (2004)
16. Rennich, S., CUDA C/C++ Streams and Concurrency. https://developer.download.nvidia.com/CUDA/training/StreamsAndConcurrencyWebinar.pdf. Accessed 16 Aug 2020

Pipelined Serial Register Renaming

Dejan Spasov[(✉)]

Faculty of Computer Science and Engineering, Skopje, Republic of North Macedonia
dejan.spasov@finki.ukim.mk

Abstract. Superscalar microarchitectures include register renaming units where architectural registers are renamed to physical registers. Modern renaming units are required to rename more than one instruction per clock cycle. Conventional renaming units use parallel circuits to simultaneously rename more than one instruction. We propose a serial circuit to rename more than one instruction in sequential manner. Then we propose a pipelined implementation of the serial renaming unit. We compare the proposed (pipelined) serial register renaming unit with the ordinary register renaming unit.

Keywords: Register renaming · Superscalar · Register alias table · Pipelining

1 Introduction

Modern superscalar cores are capable of processing more than one instruction per clock cycle at any stage of computation. More than one instruction per clock may be executed and results are stored in physical registers of the core. The core may include a large number of physical registers to eliminate false Write-After-Read (WAR) and Write-After-Write (WAW) data dependencies among instructions and to allow instructions to be executed out of program order. This approach attains higher Instructions-Per-Cycle (IPC) rates but requires register renaming unit to maintain true Read-After-Write (RAW) data dependencies between instructions waiting for execution and results stored in the physical registers [1].

Figure 1 shows the simplified microarchitecture of a core. The core includes fetch and decode unit, renaming unit, execution engine, reorder buffer, and one or more physical register files [2]. The core may include other components and interfaces not shown to simplify the presentation. The core may support multiple instruction issue, in-order and out-of-order execution, and simultaneous multi-threading. In simultaneous multi-threading instructions from one or more threads may simultaneously be processed.

The fetch and decode unit is an in-order front end unit that is configured to fetch instructions from a memory or L1 cache. Instructions may be from any instruction set architecture, e.g. PowerPC™, ARM™, SPARC™, x86™, etc. Fetched instructions consist of source and destination operands. Source and destination operands are identified with the architectural registers that are defined in the instruction set architecture.

The fetch and decode unit is configured to output a group of n instructions denoted with I(1) to I(n). Outputted instructions are considered to be in program order where

© Springer Nature Switzerland AG 2020
V. Dimitrova and I. Dimitrovski (Eds.): ICT Innovations 2020, CCIS 1316, pp. 153–161, 2020.
https://doi.org/10.1007/978-3-030-62098-1_13

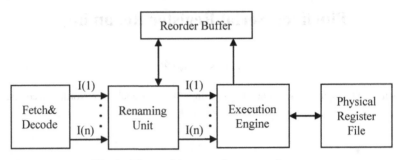

Fig. 1. Microarchitecture of a superscalar core

I(1) is the oldest instruction in the group. Source operands of instructions consumers are equal to the destination operands of the instructions producers of the result.

For each instruction with destination operand, the renaming unit is configured to allocate a physical register where the result will be stored. The renaming unit is configured to rename source operands of instructions consumers of a result to the physical register allocated to the instruction producer of the result. A source operand of an instruction is renamed to the physical register most recently allocated to instruction with destination operand equal to the source operand.

The back-end execution engine may include any number and type of execution units, e.g. integer unit, floating-point unit, load/store unit, a branch unit etc., configured to execute instructions. Instructions may be executed in-order or out-of-order. One or more reservation stations may be coupled to the execution units to host instructions waiting for execution. A dispatch unit may be included to dispatch incoming instructions to the appropriate execution units.

The core may include a reorder buffer and a physical register file to support out-of-order execution of instructions. The reorder buffer maintains the program order of the speculatively executed instructions. For each instruction, the reorder buffer allocates an entry. Instructions enter at one end of the buffer and exit (retire) at the other end. Instructions may be executed out of program order but retired in program order. Reorder buffer entries are organized as a circular buffer. In circular buffer adjacent in program order instructions allocate entries with adjacent addresses. A large number of physical registers may be organized in a physical register file. Physical registers of the core store speculative results and architecturally visible results. The large number of physical registers allows instructions to be executed speculatively and out-of-order, while false data dependencies are eliminated, and true data dependencies are maintained. The physical register file may employ stand-alone addressing scheme or may adopt the addressing scheme of the reorder buffer.

Ordinary renaming units include n parallel circuits to simultaneously rename n instructions I(1) - I(n). In this paper, we propose a pipelined serial circuit to rename the n instructions I(1) - I(n) in a sequential manner.

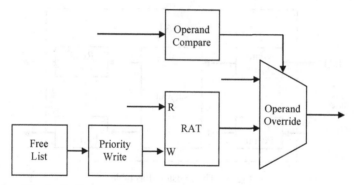

Fig. 2. Ordinary register renaming unit of a scalar microprocessor

2 Ordinary Register Renaming Unit

Register renaming unit is the place where the source and the destination operands of instructions are renamed to physical registers. Figure 2 outlines the ordinary register renaming unit. The register renaming unit is configured to rename in parallel source and destination operands in a renaming group of two or more instructions I(1) - I(n). The renaming unit includes a register alias table (RAT), an operand compare unit, an operand override unit, a priority write unit, and a free list. In cores that implement simultaneous multithreading, a separate register renaming unit is needed for each thread.

The free list maintains a list of physical registers that may be allocated to instructions with destination operands. The free list contains circuitry such that for each instruction with destination operand, the free list is configured to allocate a physical register. This way destination operands are renamed to physical registers and false WAR and WAW data dependencies among instructions are removed. When an instruction with destination operand retires, the physical register that was allocated to the instruction is added to the free list.

Source and destination operands are architectural registers. A source operand of an instruction consumer of a result is equal to the destination operand of the instruction producer of the result. When renaming source operands of instructions I(1) - I(n), an instruction I(i) may be RAW data-dependent on another instruction from the same group (inter-group dependency) or from another group (intra-group dependency).

The register alias table (RAT) maintains architectural to physical register mappings from a prior group of instructions. Intra-group dependencies are resolved by reading the RAT at indexes provided by the source operands. The operand compare unit performs tests for inter-group dependencies. The operand-override unit maintains inter-group data dependencies. The priority write unit handles priority writes to the mapping table.

The RAT may be implemented as content addressable memory (CAM). In CAM implementation RAT entries are indexed with the physical register [3, 4]. The RAT may be implemented as SRAM. In SRAM implementation RAT entries are indexed with the architectural registers [5–7]. Figure 3 shows the SRAM implementation of the register alias table. The register alias table includes one read port for each source operand and one write port for each destination operand.

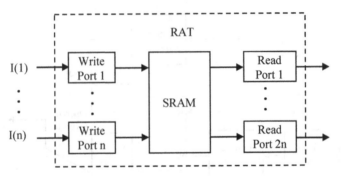

Fig. 3. The register alias table

Consider a renaming group of n instructions I(1), I(2), ..., I(n) to be provided to the register renaming unit on Fig. 2. Instructions are in program order such that instruction I(i) is older than instruction I(j) if i < j. Each instruction I(i) may be considered to include a destination operand DOP(i) and one or two source operands SOP(i). For each instruction I(i) with destination operand DOP(i), the free list is configured to allocate a physical register PR(i). The RAT includes 2n read ports coupled to receive source operands of the instructions. For each source operand SOP(i), the RAT is configured to output physical register stored at index SOP(i). The operand compare unit is configured to compare each source operand SOP(i) with destination operands of older instructions DOP(1), ..., DOP(i − 1). If a match is found, e.g. SOP(i) = DOP(j), the operand override unit is configured to output PR(j); thus, maintaining the inter-group data dependencies when renaming the source operands SOP(i). If a match is not found the operand override unit is configured to output physical register stored in the RAT at index SOP(i); thus, maintaining the intra-group data dependencies.

The priority write unit is configured to compare each destination operand DOP(i) with destination operands of younger instructions DOP(i + 1), ..., DOP(n). PR(i) may be stored in the mapping table at index DOP(i), only if a match is not found.

For a group of n instructions, ordinary register renaming unit may include n write ports and 2n (or even 3n) read ports to the RAT. One read port is needed for each source operand and in certain implementations one read port is used to read the physical register that is overwritten by the destination operand. Each read port is coupled to receive a source operand from the group. Each write port is coupled to receive a destination operand from the group. The RAT may be implemented as a multi-ported SRAM. The die area of multi-ported SRAM grows quadratically with the number of ports. Power consumption and power density of the SRAM memory are proportional to the die area [8].

The priority write unit may include ~n^2 comparators to compare each destination operand with destination operands of older instructions from the group. The operand compare unit may include ~n^2 comparators to compare each source operand with destination operands of older instructions in the group. Therefore, we can conclude that die area, wiring complexity, power consumption, and power density of the ordinary register renaming unit grow quadratically with respect to the number of instructions n in the renaming group.

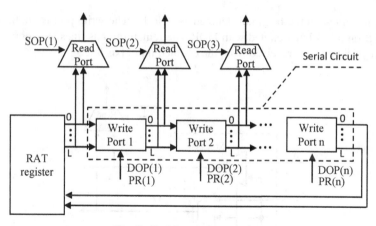

Fig. 4. Serial register renaming

3 Serial Register Renaming Unit

Figure 4 shows the proposed serial register renaming unit. The serial renaming unit uses a RAT register instead of a RAT table to maintain architectural to physical register mappings. The RAT register uses one write port and one read port to maintain architectural to physical register mappings from prior groups of instructions. The RAT register consists of fields such that for each architectural register there is a unique field. Fields of the RAT register store physical registers most recently allocated to instructions, from prior renaming groups, with destination operands one-to-one associated with the fields.

The proposed renaming unit includes a serial circuit that consists of n write ports coupled in a chain. Architectural to physical register mappings propagate from the RAT register through the chain of write ports over bus lines denoted with the architectural registers 0 to L. A bus line denoted with the architectural register I may be considered to propagate a physical register allocated to an instruction with destination operand I. Each write port is coupled to receive destination operand DOP(i) and allocated physical register PR(i) of an instruction I(i). The write port is configured to update architectural to physical register mappings with the mapping DOP(i) to PR(i). The first write port, coupled to the mapping table, is configured to output PR(1) on a bus line denoted with DOP(1). The second write port, coupled to the first write port, is configured to output PR(2) on a bus line denoted with DOP(2), etc. The chain of write ports sequentially in program order inserts allocated physical registers PR(1), PR(2), ..., PR(n) on bus lines denoted with DOP(1), DOP(2), ..., DOP(n), respectively. The last write port is coupled to the RAT register to write updated architectural to physical register mappings.

Figure 5 shows the write port of the proposed renaming unit. The write port is coupled to receive architectural to physical register mappings on the bus lines denoted with 0 to L. Bus lines 0 to L are terminated at 2-in-1 multiplexers. The other input on the 2-in-1 multiplexers is coupled to receive a physical register PR allocated to an instruction with destination operand DOP. A decoder is coupled to receive destination operand DOP. Output signal lines of the decoder are coupled as selection control to the 2-in-1 multiplexers. Multiplexers are configured to output architectural to physical register mappings on the bus lines denoted as $0'$ to L'. The write port is configured to output PR on a bus line denoted with DOP. The remaining bus lines are configured to output physical register received from the coupled input bus line.

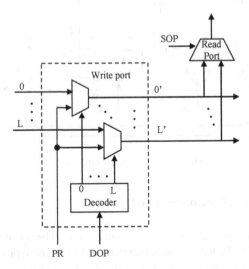

Fig. 5. Write port of the proposed renaming unit

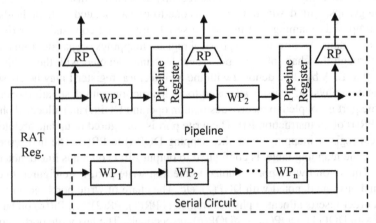

Fig. 6. Pipelined serial register renaming unit

A read port, basically a mux, is coupled to the output of the write port. Source operand SOP of an instruction is provided as selection control to the read port. The read port is configured to output physical registers propagating on the bus line denoted with SOP; thus, renaming the source operand SOP to a physical register.

The proposed serial register rename unit does not need the priority write unit, operand compare unit, and the operand override unit. The SRAM cells of the RAT are operated as register with one read port and one write port. Using a register instead of a table reduces wiring complexity (bit lines), the number of circuitries for pre-charge, the number of sense amplifiers, and the number of write drivers. Therefore, none of the building blocks of the proposed rename unit is with quadratic complexity.

4   Pipelined Serial Register Renaming Unit

One advantage of the serial renaming unit is the ability to implement pipelining in the renaming process. Figure 6 shows the pipelined implementation of the proposed serial register renaming unit. The RAT register in the pipelined implementation includes one write port and two read ports. A serial circuit made of n write ports WP_1 to WP_n is coupled to read the content of the RAT register to update the architectural to physical register mappings and to write the updated architectural to physical register mappings in the RAT register for the following group of instructions. The pipeline consists of a chain of write ports intersected with pipeline registers. Read ports are coupled to the write ports in the pipeline. In one clock cycle, the serial circuit is configured to update the RAT register for the next renaming group. The pipeline is configured to rename source operands in the subsequent cycles. The pipelined renaming unit includes 2 read ports and 1 write port and at most n pipeline registers. Pipeline registers are with same size as the RAT register. Thus, it is obvious that the pipelined renaming unit is with the greatest hardware complexity. However, it may offer the best performance compared to the other renaming techniques in terms of attainable clock frequency.

5   Experimental Results

We have designed a Verilog description of the ordinary register renaming unit (Fig. 2), the serial register renaming unit (Fig. 4), and the pipelined register renaming unit (Fig. 6). Then, we have tested the renaming units using the Quartus software targeting the Cyclone FPGA family [12]. We have observed the f_{max} obtained from the timing analyzer for various sizes of the renaming group. Results are given in Fig. 7.

Figure 7 shows maximum attainable clock frequency (f_{max}) that can be achieved for all three renaming units for various sizes of the renaming group. Modern state-of-the-art processors implement renaming groups with 5 or 6 instructions. Overall, the pipelined register renaming shows the best results and expectably serial register renaming performs the worst. The reason for the underperforming of the serial register renaming is the length of the critical path from the RAT register to the read port for the youngest instruction (Fig. 4). From Fig. 7, one observes that ordinary register renaming attains higher operating frequency than the serial register renaming unit for renaming groups with more than 2 instructions. Pipelined renaming shows higher attainable frequencies that the ordinary register renaming for renaming groups with less than 9 instructions. In our simulations we have used the largest possible number pipeline registers to observe maximum improvement. We should note that this number of pipeline registers may be reduced at the design stage of the microprocessor.

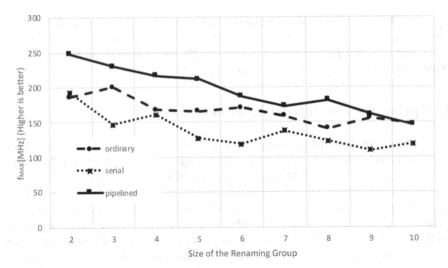

Fig. 7. Maximum attainable clock frequency for various sizes of the renaming group

6 Related Work

Complexity, power density, and power consumption are major weaknesses of the register renaming units. In order to overcome these weaknesses, several techniques have been proposed. In general, these techniques focus on reducing the number of ports to the mapping table and adding additional memory (caches) to relieve the pressure on the mapping table.

In [9], A. Moshovos studies the average number of operands in the SPEC2000 benchmark programs and in the multimedia applications. The author proposes a reduction on the number of read ports at the cost of negligent slowdown of the instruction-per-cycle performance.

In [10], G. Kucuk, et al., propose a mechanism to reduce power consumption. In particular, if a source operand depends on a destination operand from the same renaming group, read port for that source operand is turned off. Moreover, authors propose buffering of the renamed destination operands in a small number of latches; thus, reducing the pressure on the mapping table.

In [8], E. Safi, et al. propose two stage pipelined register renaming technique. Priority write unit, operand compare unit and operand override unit are placed in one of the pipeline stages, while reading and writing to the mapping table in the other pipeline stage.

In [11], S. Petit, et al. use RAM-based and CAM-based tables to perform what they call hybrid renaming at the expense of increased power dissipation.

In [12], S. Vajapeyam, et al. and in [13] R. Shioya, et al. propose techniques for caching renamed operands.

7   Conclusion

Modern state-of-the-art microprocessor architectures are designed to rename about 5 to 6 instructions. We have proposed pipelined register renaming units that can rename 5 or 6 instructions, while achieving higher clock frequencies in the Cyclone FPGA families. The proposed register renaming unit provides opportunity to the designers of the microprocessors to increase the clock frequency or to increase the size n of the renaming group. Increased clock frequency or increased the size n of the renaming group improve performance of the microprocessors.

References

1. Hennessy, J.L., Patterson, D.A.: Computer Architecture: A Quantitative Approach, 5th edn. Elsevier, Netherlands (2011)
2. Smith, J.E., Sohi, G.S.: The microarchitecture of superscalar processors. Proc. IEEE **83**(12), 1609–1624 (1995)
3. Kessler, R.: The Alpha 21264 microprocessor. IEEE Micro **19**(2), 24–36 (1999)
4. Buti, T.N., et al.: Organization and implementation of the register-renaming mapper for out-of-order IBM POWER4 processors. IBM J. Res. Dev. **49**(1), 167–188 (2005)
5. Tomasulo, R.M.: An efficient algorithm for exploiting multiple arithmetic units. IBM J. Res. Dev. **11**(1), 25–33 (1967)
6. Hinton, G., et. al.: The microarchitecture of the Pentium® 4 processor. Intel Technol. J. (2001)
7. Sadasivam, S.K., Thompto, B.W., Kalla, R., Starke, W.J.: IBM Power9 processor architecture. IEEE Micro **37**(2), 40–51 (2017)
8. Safi, E., Moshovos, A., Veneris, A.: Two-stage, pipelined register renaming. IEEE Trans. Very Large-Scale Integr. (VLSI) Syst. **19**(10), 1926–1931 (2011)
9. Moshovos, A.: Power-aware register renaming. Technical report, Computer Engineering Group, University of Toronto (2002)
10. Kucuk, G., Ergin, O., Ponomarev, D., Ghose, K.: Reducing power dissipation of register alias tables in high-performance processors. IEE Proc. Comput. Digit. Tech. **152**(6), 739–746 (2005)
11. Petit, S., Ubal, R., Sahuquillo, J., López, P.: Efficient register renaming and recovery for high-performance processors. IEEE Trans. Very Large-Scale Integr. (VLSI) Syst. **22**(7), 1506–1514 (2014)
12. Vajapeyam, S., Mitra, T.: Improving superscalar instruction dispatch and issue by exploiting dynamic code sequences. ACM SIGARCH Comput. Archit. News **25**(2), 1–12 (1997)
13. Shioya, R., Ando, H.: Energy efficiency improvement of renamed trace cache through the reduction of dependent path length. In: Proceedings of the IEEE 32nd International Conference on Computer Design (ICCD), New York, NY, USA, pp. 416–423. IEEE (2014)

Fast Decoding with Cryptcodes for Burst Errors

Aleksandra Popovska-Mitrovikj$^{(\boxtimes)}$, Verica Bakeva, and Daniela Mechkaroska

Faculty of Computer Science and Engineering, Ss. Cyril and Methodius University, Skopje, Republic of North Macedonia
{aleksandra.popovska.mitrovikj,verica.bakeva,
daniela.mechkaroska}@finki.ukim.mk

Abstract. Random Codes Based on Quasigroups (RCBQ) are crypt-codes that provide a correction of transmission errors and an information security, all with one algorithm. Standard algorithm, Cut-Decoding and 4-Sets-Cut-Decoding algorithms are different versions of RCBQ and they are proposed elsewhere. The decoding in all these algorithms is list decoding, so the speed of the decoding depends on the list size. In order to decrease the list size, Fast-Cut-Decoding and Fast-4-Sets-Cut-Decoding algorithms are proposed elsewhere and they improve performances of these codes for transmission through a Gaussian channel. Here, we propose a new modification of these algorithms to improve their properties for transmission through a burst channels.

Keywords: Cryptcoding · Burst errors · Gilbert-Elliott channel · SNR · Quasigroup

1 Introduction

A correct transmission through a noisy channel and providing the information security of transmitted data are two most important things in communication systems. Usually, this is achieved by using two algorithms: one for correction of errors and another for obtaining information security [4,13,14]. The concept of cryptcoding merges these two algorithms in one by using a cryptographic algorithm during the encoding/decoding process.

Random Codes Based on Quasigroups (RCBQ) are cryptcodes. Therefore, they allow not only correction of certain amount of errors in the input data, but they also provide an information security, all built in one algorithm. For the first time, they are proposed from Gligoroski and et al. in [1]. We named this coding/decoding algorithm as Standard algorithm. The main problem of these codes is the speed of decoding. Namely, the decoding in RCBQ is actually the list decoding, so the speed of this process depends on the list size. For improving the performances of these codes, Cut-Decoding and 4-Sets-Cut-Decoding algorithms are proposed in [9,12]. In the decoding process of these codes, three types of errors appear: *more-candidate-error*, *null-error* and *undetected-error*.

© Springer Nature Switzerland AG 2020
V. Dimitrova and I. Dimitrovski (Eds.): ICT Innovations 2020, CCIS 1316, pp. 162–173, 2020.
https://doi.org/10.1007/978-3-030-62098-1_14

More-candidate-errors can occur even all bits in the message are correctly transmitted. Therefore, the packet-error and bit-error probabilities can be positive for very small bit-error probability in the noisy channel. In order to eliminate this problem, in paper [8], we defined decoding algorithms called Fast-Cut-Decoding and Fast-4-Sets-Cut-Decoding algorithms. These algorithms enable more efficient and faster decoding, especially for transmission through a low noise channel. In [8], the performances of these algorithms for transmission through a Gaussian channel are considered.

On the other side, in experiments with burst channels, Cut-Decoding and 4-Sets-Cut-Decoding algorithms do not give good results. Therefore, in [6], we proposed algorithms for coding/decoding when burst channels are used. These algorithms are called Burst-Cut-Decoding and Burst-4-Sets-Cut-Decoding algorithms.

Our goal in this paper is to adopt Fast-Cut-Decoding and Fast-4-Sets-Cut-Decoding algorithms for transmission through burst channels. For that purpose, here we propose two new algorithms called FastB-Cut-Decoding and FastB-4-Sets-Cut-Decoding algorithms.

The paper is organized as follows. In Sect. 2, we briefly describe Standard, Cut-Decoding and 4-Sets-Cut-Decoding algorithms for RCBQ. The description of new algorithms (FastB-Cut-Decoding and FastB-4-Sets-Cut-Decoding) is given in Sect. 3. In Sect. 4, we present several experimental results obtained with these new algorithms. We analyze the results for packet-error, bit-error probabilities and decoding speed when messages are transmitted through a burst channel. The burst errors are simulated using Gilbert-Elliott model with Gaussian channels. Also, we compare these results with the results obtained with the old algorithms for burst channels. At the end, we give some conclusions for presented results.

2 Description of Standard, Cut-Decoding and 4-Sets-Cut-Decoding Algorithms

In the coding/decoding algorithms of RCBQ encryption/decryption algorithms from the implementation of Totally Asynchronous Stream Cipher (TASC) by quasigroup string transformation [2] are used. These cryptographic algorithms use the alphabet Q and a quasigroup operation $*$ on Q together with its parastrophe $\backslash$. In our experiments with RCBQ we use the alphabet of 4-bit symbols (nibbles).

The notions of quasigroups and quasigroup string transformations are given in the previous papers for these codes [7,9–11]. Here, we use the same terminology and notations as there.

2.1 Coding Algorithms

In Standard algorithm of RCBQ [1], the process of coding is as follows. First the message M of $N_{block} = 4l$ bits (l nibbles) is extended to message L of

$N = 4m$ bits (m nibbles) by adding redundant zero symbols according to a chosen pattern. Then, we choose a key $k = k_1k_2...k_n \in Q^n$ and by applying the encryption algorithm of TASC (given in Fig. 1) on the message L, we obtain the codeword $C = C_1C_2...C_m$ ($C_i \in Q$) for the message M. So, we have (N_{block}, N) code with rate $R = N_{block}/N$.

Encryption	Decryption
Input: Key $k = k_1k_2\ldots k_n$ and	**Input:** The pair
$L = L_1L_2\ldots L_m$	$(a_1a_2\ldots a_r, k_1k_2\ldots k_n)$
Output: codeword	**Output:** The pair
$C = C_1C_2...C_m$	$(c_1c_2\ldots c_r, K_1K_2\ldots K_n)$
For $j = 1$ to m	For $i = 1$ to n
$\quad X \leftarrow L_j;$	$\quad K_i \leftarrow k_i;$
$\quad T \leftarrow 0;$	For $j = 0$ to $r - 1$
$\quad$ For $i = 1$ to n	$\quad X, T \leftarrow a_{j+1};$
$\quad\quad X \leftarrow k_i * X;$	$\quad temp \leftarrow K_n;$
$\quad\quad T \leftarrow T \oplus X;$	$\quad$ For $i = n$ to 2
$\quad\quad k_i \leftarrow X;$	$\quad\quad X \leftarrow temp \setminus X;$
$\quad k_n \leftarrow T$	$\quad\quad T \leftarrow T \oplus X;$
$\quad$ **Output:** $C_j \leftarrow X$	$\quad\quad temp \leftarrow K_{i-1};$
	$\quad\quad K_{i-1} \leftarrow X;$
	$\quad X \leftarrow temp \setminus X;$
	$\quad K_n \leftarrow T;$
	$\quad c_{j+1} \leftarrow X;$
	Output: $(c_1c_2\ldots c_r, K_1K_2\ldots K_n)$

Fig. 1. Algorithms for encryption and decryption

In Cut-Decoding algorithm, in order to obtain (N_{block}, N) code with rate R we use two ($N_{block}, N/2$) codes with rate $2R$ for coding/decoding the same message of N_{block} bits. In the process of coding we apply the encryption algorithm (given in Fig. 1) two times, on the same redundant message L using different parameters (different keys or quasigroups). The codeword of the message is a concatenation of two codewords of $N/2$ bits. In 4-Sets-Cut-Decoding algorithm we use four ($N_{block}, N/4$) codes with rate $4R$, on the same way as in Cut-Decoding algorithm and the codeword of the message is a concatenation of four codewords of $N/4$ bits.

2.2 Decoding Algorithms

In Standard RCBQ, after transmission of the codeword C through a noisy channel, we divide the received message D in s blocks of r nibbles ($D = D^{(1)}$ $D^{(2)} \ldots D^{(s)}$). Then we choose an integer B_{max} which is the assumed maximum number of bit errors that occur in a block during transmission. In each iteration we generate the sets $H_i, i = 1, 2, \ldots, s$ of all strings with r nibbles that are at Hamming's distance $\leq B_{max}$ from the corresponding block $D^{(i)}$ and the decoding candidate sets $S_0, S_1, S_2, \ldots, S_s$. These sets are defined iteratively, and $S_0 = (k_1 \ldots k_n; \lambda)$, where λ is the empty sequence. In the ith iteration, we form the set S_i of pairs ($\delta, w_1w_2 \ldots w_{4ri}$) by using the sets S_{i-1} and

H_i as follows (w_j are bits). For each element $\alpha \in H_i$ and each $(\beta, w_1 w_2 \dots w_{4r(i-1)}) \in S_{i-1}$, we apply the decryption algorithm given in Fig. 1 with input (α, β). If the output is the pair (γ, δ) and if the sequence γ has redundant zeros at the right positions (according to the chosen pattern), then the pair $(\delta, w_1 w_2 \dots w_{4r(i-1)} c_1 c_2 \dots c_r) \equiv (\delta, w_1 w_2 \dots w_{4ri})$ $(c_i \in Q)$ is an element of S_i.

In Cut-Decoding algorithm, after transmission through a noisy channel, we divide the outgoing message $D = D^{(1)} D^{(2)} \dots D^{(s)}$ in two messages (D_1 and D_2) with equal lengths and we decode them parallel with the corresponding parameters. In this decoding algorithm in each iteration we reduce the number of elements in the decoding candidate sets on the following way. Let $S_i^{(1)}$ and $S_i^{(2)}$ be the decoding candidate sets obtained in the ith iteration of two parallel decoding processes, $i = 1, \dots, s/2$. We eliminate from $S_i^{(1)}$ all elements whose second part does not match with the second part of an element in $S_i^{(2)}$, and vice versa. In the $(i+1)^{th}$ iteration the both processes use the corresponding reduced sets. On this way, the size of the lists (decoding candidate sets) becomes smaller.

In 4-Sets-Cut-Decoding algorithm after transmitting through a noisy channel, we divide the outgoing message $D = D^{(1)} D^{(2)} \dots D^{(s)}$ in four messages D^1, D^2, D^3 and D^4 with equal lengths and we decode them parallelly with the corresponding parameters. Similarly, as in Cut-Decoding algorithm, in each iteration of the decoding process we reduce the decoding candidate sets obtained in the four decoding processes. In [12], the authors proposed 4 different versions of decoding with 4-Sets-Cut-Decoding algorithm. They differ in the way of reduction of the decoding candidate sets. In the experiments presented in this paper we use only the version that gave the best results. In this version of the algorithm the four decoding candidate are reduced as follows. Let $S_i^{(1)}$, $S_i^{(2)}$, $S_i^{(3)}$ and $S_i^{(4)}$ be the decoding candidate sets obtained in the ith iteration of four parallel decoding processes, $i = 1, \dots, s/4$. Let $V_1 = \{w_1 w_2 \dots w_{r \cdot a \cdot i} | (\delta, w_1 w_2 \dots w_{r \cdot a \cdot i}) \in S_i^{(1)}\}$, $\dots$, $V_4 = \{w_1 w_2 \dots w_{r \cdot a \cdot i} | (\delta, w_1 w_2 \dots w_{r \cdot a \cdot i}) \in S_i^{(4)}\}$ and $V = V_1 \cap V_2 \cap V_3 \cap V_4$. If $V = \emptyset$ then $V = (V_1 \cap V_2 \cap V_3) \cup (V_1 \cap V_2 \cap V_4) \cup (V_1 \cap V_3 \cap V_4) \cup (V_2 \cap V_3 \cap V_4)$. Before the next iteration we eliminate from $S_i^{(j)}$ all elements whose second part is not in V, $j = 1, 2, 3, 4$. With this elimination, as in Cut-Decoding algorithm, we decrease the length of the lists.

After the last iteration, if all reduced sets (two in Cut-Decoding algorithm, four in 4-Sets-Cut-Decoding) have only one element with a same second component then this component is the decoded message L. In this case, we say that we have a *successful decoding*. If the decoded message is not the correct one then we have an *undetected-error*. If the reduced sets obtained in the last iteration have more than one element then we have a *more-candidate-error*. In this case we randomly select a message from the reduced sets in the last iteration and we take this message as the decoded message. If in some iteration all decoding candidate sets are empty, then the process will finish (we say that a *null-error* appears). But, if we obtain at least one nonempty decoding candidate set in an iteration then the decoding continues with the nonempty sets (the reduced sets are obtained by intersection of the non-empty sets only).

3   Fast Algorithms for Burst Channels

As we mentioned previously, decoding with Standard, Cut-Decoding and 4-Sets-Cut-Decoding algorithms is actually list decoding. Therefore, the speed of decoding process depends on the list size (a shorter list gives faster decoding). In all algorithms, the list size depends on B_{max} (the maximal assumed number of bit errors in a block). For smaller values of B_{max}, shorter lists are obtained. But, the number of errors during transmission of a block is not known in advance. If this number of errors is larger than assumed number of bit errors B_{max} in a block, the errors will not be corrected. On the other side, if B_{max} is too large, we have long lists and the process of decoding is too slow. Also, larger value of B_{max} can lead to ending of the decoding process with a *more-candidate-error*. In this case, the correct message will be in the list of the last iteration, if there are no more than B_{max} errors during transmission. Therefore, with all decoding algorithms for RCBQ, *more-candidate-errors* can be obtained, although the bit-error probability of the channel is small and the number of bit errors in a block is not greater than B_{max} (or no errors during transmission).

In order to solve this problem, in [8], we proposed a modification of Cut-Decoding and 4-Sets-Cut-Decoding algorithms, called Fast-Cut-Decoding and Fast-4-Sets-Cut-Decoding algorithms. There, instead of a fixed value B_{max}, the decoding process in the both algorithms starts with $B_{max} = 1$. If successful decoding is obtained, the procedure is done. If not, the value of B_{max} is increased with 1 and the decoding process is repeated with the new value of B_{max}, etc. The decoding finishes with $B_{max} = 4$ (for rate 1/4) or with $B_{max} = 5$ (for rate 1/8). These algorithms try to decode the message using the shorter lists and in the case of successful decoding with a small value of B_{max} ($B_{max} < 4$), long lists are avoid and the decoding is faster. Also, the number of *more-candidate-errors* decreases.

On the other side, Cut-Decoding and 4-Sets-Cut-Decoding algorithms do not give a good results when transmission is through a burst channel. Therefore, in [6], we proposed new algorithms called Burst-Cut-Decoding and Burst-4-Sets-Cut-Decoding algorithms. In order to handling burst errors in a communication system, in these algorithms, we include an interleaver in coding algorithm and the corresponding deinterleaver in the decoding algorithm. Namely, in the process of coding before the concatenation of two (or four) codewords we apply the interleaving on each codeword, separately. The interleaver rearranges (by rows) m nibbles of a codeword in a matrix of order $(m/k) \times k$. The output of the interleaver is a mixed message obtained reading the matrix by columns. Then, after transmission of a concatenated message through a burst channel we divide the outgoing message D in two (or four) messages with equal length and before the parallel decoding we apply deinterleaving on each message, separately. The coding/decoding process with these algorithms is schematically presented on Fig. 2.

In this paper, we want to adopt fast algorithms of RCBQ for transmission through a burst channel. So, we combine the previous two ideas in two new algorithms for coding/decoding called FastB-Cut-Decoding and FastB-4-Sets-

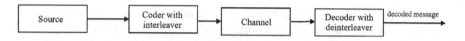

Fig. 2. Coding/decoding process in the new algorithms

Cut-Decoding algorithms. Namely, at first we apply one of the coding algorithm (Fast-Cut-Decoding or Fast-4-Sets-Cut-Decoding algorithms) on the original message. After that we apply interleaving on the obtained codewords (before concatenation). Interleaved message is transmitted through a noisy burst channel. On the received messages on the output of the channel, after dividing in two (or four) parts we apply deinterleaving of each part and then all parts are decoded using the appropriate decoding algorithm.

4   Experimental Results

In this section, we present experimental results obtained with FastB-Cut-Decoding and FastB-4-Sets-Cut-Decoding algorithms for rate $R = 1/4$ and $R = 1/8$, for transmission through a burst channel. The simulation of burst errors is made using Gilbert-Elliott model with Gaussian channels. This model is based on a Markov chain with two states G (good or gap) and B (bad or burst). In good state the probability for incorrect transmission of a bit is small, and in bad state this probability is large. The model is defined with the transmission probability P_{BB} from bad to bad state and probability P_{GG} from good to good state [3]. In our experiments we use Gilbert-Elliott model where in both states the channel is a Gaussian and SNR_G in a good state is high and SNR_B in a bad state is low.

We compare results obtained with FastB-Cut-Decoding and FastB-4-Sets-Cut-Decoding algorithms with the corresponding results for Burst-Cut-Decoding and Burst-4-Sets-Cut-Decoding algorithms. Also, in order to show the efficiency of Fast algorithms we present percentages of messages which decoding finished with $B_{max} = 1, 2, 3, 4$ or 5.

In the experiments we use the following code parameters.

- For code $(72, 288)$ in Burst-Cut-Decoding and FastB-Cut-Decoding algorithm, the parameters are:
  - redundancy pattern: 1100 1110 1100 1100 1110 1100 1100 1100 0000 for rate 1/2 and two different keys of 10 nibbles.
- For code $(72, 576)$, the code parameters are:
  - in Burst-Cut-Decoding and FastB-Cut-Decoding - redundancy pattern: 1100 1100 1000 00001100 1000 1000 0000 1100 1100 1000 0000 1100 1000 1000 0000 0000 0000, for rate 1/4 and two different keys of 10 nibbles,
  - in Burst-4-Sets-Cut-Decoding and FastB-4-Sets-Cut-Decoding - redundancy pattern: 1100 1110 1100 1100 1110 1100 1100 1100 0000 for rate 1/2 and four different keys of 10 nibbles.

In all experiments we use the same quasigroup on Q given in Table 1.

In the experiments with Burst-Cut-Decoding for the code $(72, 288)$ we use $B_{max} = 4$, and in the experiments with FastB-Cut-Decoding algorithm the maximum value of B_{max} is 4. For the code $(72, 576)$ in Burst-Cut-Decoding and Burst-4-Sets-Cut-Decoding algorithms we use $B_{max} = 5$, and in FastB-Cut-Decoding and FastB-4-Sets-Cut-Decoding algorithms the maximum value of B_{max} is 5.

Table 1. Quasigroup of order 16 used in the experiments

*	0	1	2	3	4	5	6	7	8	9	a	b	c	d	e	f
0	3	c	2	5	f	7	6	1	0	b	d	e	8	4	9	a
1	0	3	9	d	8	1	7	b	6	5	2	a	c	f	e	4
2	1	0	e	c	4	5	f	9	d	3	6	7	a	8	b	2
3	6	b	f	1	9	4	e	a	3	7	8	0	2	c	d	5
4	4	5	0	7	6	b	9	3	f	2	a	8	d	e	c	1
5	f	a	1	0	e	2	4	c	7	d	3	b	5	9	8	6
6	2	f	a	3	c	8	d	0	b	e	9	4	6	1	5	7
7	e	9	c	a	1	d	8	6	5	f	b	2	4	0	7	3
8	c	7	6	2	a	f	b	5	1	0	4	9	e	d	3	8
9	b	e	4	9	d	3	1	f	8	c	5	6	7	a	2	0
a	9	4	d	8	0	6	5	7	e	1	f	3	b	2	a	c
b	7	8	5	e	2	a	3	4	c	6	0	d	f	b	1	9
c	5	2	b	6	7	9	0	e	a	8	c	f	1	3	4	d
d	a	6	8	4	3	e	c	d	2	9	1	5	0	7	f	b
e	d	1	3	f	b	0	2	8	4	a	7	c	9	5	6	e
f	8	d	7	b	5	c	a	2	9	4	e	1	3	6	0	f

In all experiments the value of SNR in the good state is $SNR_G = 4$. We made experiments for different values of SNR in the bad state $SNR_B \in \{-3, -2, -1\}$ and for the following combinations of transition probabilities from good to good state P_{GG} and from bad to bad state P_{BB} in the Gilbert-Elliott model:

- $P_{GG} = 0.8$ and $P_{BB} = 0.8$
- $P_{GG} = 0.5$ and $P_{BB} = 0.5$
- $P_{GG} = 0.2$ and $P_{BB} = 0.8$
- $P_{GG} = 0.8$ and $P_{BB} = 0.2$

In Table 2, we give experimental results for bit-error probabilities BER and packet-error probabilities PER for the code $(72, 288)$. With BER_{b-cut} and PER_{b-cut} we denote probabilities obtained with Burst-Cut-Decoding algorithm, and with BER_{f-cut} and PER_{f-cut} probabilities obtained with FastB-Cut-Decoding algorithm. The results for BER_{b-cut} and PER_{b-cut} for Burst-Cut-Decoding algorithm are published in [6].

Table 2. Experimental results for code $(72, 288)$

SNR_B	PER_{f-cut}	PER_{b-cut}	BER_{f-cut}	BER_{b-cut}
	$P_{GG} = 0.8$	$P_{BB} = 0.8$		
-3	0.27657	0.32366	0.20524	0.24605
-2	0.15431	0.17727	0.11143	0.13127
-1	0.06588	0.08179	0.04531	0.05911
	$P_{GG} = 0.5$	$P_{BB} = 0.5$		
-3	0.20838	0.23135	0.14832	0.16945
-2	0.10023	0.12348	0.06993	0.08867
-1	0.04097	0.05609	0.02839	0.03993
	$P_{GG} = 0.2$	$P_{BB} = 0.8$		
-3	0.57402	0.57783	0.43486	0.44170
-2	0.35088	0.35419	0.25447	0.26129
-1	0.15315	0.16561	0.10795	0.12115
	$P_{GG} = 0.8$	$P_{BB} = 0.2$		
-3	0.02254	0.03513	0.01529	0.02449
-2	0.00994	0.01980	0.00669	0.01393
-1	0.00382	0.00986	0.00253	0.00589

Analyzing the results in Table 2, we can conclude that for all values of SNR_B and all combinations of transition probabilities results for BER_{f-cut} and PER_{f-cut} are slightly better than the corresponding results of BER_{b-cut} and PER_{b-cut}.

When decodings in FastB-Cut-Decoding algorithm end with $B_{max} = 1$ or $B_{max} = 2$ decoding with this algorithm is much faster than with Burst-Cut-Decoding. Therefore, in Table 3 we give the percentage of messages for which decoding ended with $B_{max} = 1$, $B_{max} = 2$, $B_{max} = 3$ or $B_{max} = 4$ in the experiments with FastB-Cut-Decoding algorithm. From the results given there, we can see that the percentages depend not only on the value of SNR_B but also on the transition probabilities for bad-to-bad and good-to-good state. For $P_{GG} = 0.8, P_{BB} = 0.2$ and all values of SNR_B for more than 75% of the messages decoding successfully finished with $B_{max} = 1$ or $B_{max} = 2$. For $P_{GG} = 0.8, P_{BB} = 0.8$ or $P_{GG} = 0.5, P_{BB} = 0.5$ and $SNR_B = -1$ more than 50% of the decoding successfully finished with $B_{max} = 1$ or $B_{max} = 2$. In other experiments, where the probability the channel to be in a bad state is greater we have a smaller percentages for successful decoding with $B_{max} = 1$ or $B_{max} = 2$. So, we can conclude that the new algorithm improves decoding speed of RCBQs for transmission through a Gilbert-Elliott channel with smaller probability for bad state.

In Table 4 and Table 5, we present the experimental results for code $(72, 576)$ with rate $R = 1/8$. We compare bit-error (BER) and packet-error (PER) prob-

Table 3. Percentage of messages decoded with different values of B_{max}

SNR_B	$B_{max} = 1$	$B_{max} = 2$	$B_{max} = 3$	$B_{max} = 4$
$P_{GG} = 0.8$		$P_{BB} = 0.8$		
-3	15.79%	18.23%	17.71%	48.26%
-2	20.06%	22.56%	22.34%	35.04%
-1	24.95%	29.59%	24.25%	21.21%
$P_{GG} = 0.5$		$P_{BB} = 0.5$		
-3	5.11%	20.39%	26.30%	48.19%
-2	8.92%	29.10%	30.34%	31.64%
-1	18.17%	37.82%	27.70%	16.31%
$P_{GG} = 0.2$		$P_{BB} = 0.8$		
-3	0.17%	3.28%	10.95%	85.60%
-2	0.42%	8.55%	21.45%	69.58%
-1	2.17%	20.99%	32.60%	44.24%
$P_{GG} = 0.8$		$P_{BB} = 0.2$		
-3	42.42%	32.80%	16.32%	8.46%
-2	50.97%	32.00%	12.66%	4.37%
-1	60.10%	29.56%	8.13%	2.22%

abilities obtain with Burst-Cut-Decoding, FastB-Cut-Decoding, Burst-4-Sets-Cut-Decoding and FastB-4-Sets-Cut-Decoding algorithms. With BER_{b-cut}, PER_{b-cut}, $BER_{b-4sets}$, $PER_{b-4sets}$ we denote probabilities for package- and bit-errors obtained with Burst algorithms (Burst-Cut-Decoding and Burst-4-Sets-Cut-Decoding algorithms) and with BER_{f-cut}, PER_{f-cut}, $BER_{f-4sets}$ and $PER_{f-4sets}$ the corresponding probabilities obtained with FastB algorithms (FastB-Cut-Decoding and FastB-4-Sets-Cut-Decoding).

From the results given in Table 4 and Table 5, we can conclude that for all combinations of transition probabilities and all values of SNR_B, results for BER and PER obtained with new FastB algorithms are better than the corresponding results obtained with Burst algorithms. Comparing the results of all algorithms, we can see that the best results (for all channel parameters) are obtained with FastB-4-Sets-Cut-Decoding algorithm. For $P_{GG} = 0.8, P_{BB} = 0.2$ and all considered values of SNR_B, the values of BER and PER obtained with these algorithms are equal to 0. Also, from the duration of the experiments we concluded that FastB-4-Sets-Cut-Decoding algorithm is much faster than the other three algorithms considered in this paper.

In Table 6, we give the percentage of messages which decoding with FastB-Cut-Decoding and FastB-4-Sets-Cut-Decoding algorithms ended with $B_{max} = 1$, $B_{max} = 2$, $B_{max} = 3$, $B_{max} = 4$ or $B_{max} = 5$.

From Table 6, we can see that in all experiments we obtained better percentages (larger percentages for smaller B_{max} and smaller percentages for

Table 4. Experimental results for BER for code $(72, 576)$

SNR_B	BER_{f-cut}	BER_{b-cut}	$BER_{f-4sets}$	$BER_{b-4sets}$
	$P_{GG} = 0.8$	$P_{BB} = 0.8$		
-3	0.06381	0.09182	0.01421	0.03798
-2	0.02148	0.03935	0.00322	0.01631
-1	0.00624	0.01347	0.00044	0.00518
	$P_{GG} = 0.5$	$P_{BB} = 0.5$		
-3	0.03417	0.05539	0.00588	0.02263
-2	0.01257	0.02198	0.00138	0.00853
-1	0.00250	0.00707	0.00022	0.00251
	$P_{GG} = 0.2$	$P_{BB} = 0.8$		
-3	0.16994	0.19376	0.06586	0.08592
-2	0.06408	0.08687	0.01560	0.03244
-1	0.01697	0.02736	0.00258	0.01085
	$P_{GG} = 0.8$	$P_{BB} = 0.2$		
-3	0.00200	0.00477	0	0.00225
-2	0.00063	0.00174	0	0.00091
-1	0.00013	0.00075	0	0.00026

Table 5. Experimental results for PER for code $(72, 576)$

SNR_B	PER_{f-cut}	PER_{b-cut}	$PER_{f-4sets}$	$PER_{b-4sets}$
	$P_{GG} = 0.8$	$P_{BB} = 0.8$		
-3	0.11470	0.16734	0.02657	0.06552
-2	0.04155	0.07273	0.00662	0.02931
-1	0.01202	0.02556	0.00079	0.00900
	$P_{GG} = 0.5$	$P_{BB} = 0.5$		
-3	0.06509	0.10765	0.01202	0.03910
-2	0.02304	0.04198	0.00324	0.01419
-1	0.00540	0.01375	0.00043	0.00446
	$P_{GG} = 0.2$	$P_{BB} = 0.8$		
-3	0.29551	0.35628	0.12025	0.14653
-2	0.11874	0.16381	0.02988	0.05688
-1	0.03283	0.05357	0.00490	0.01879
	$P_{GG} = 0.8$	$P_{BB} = 0.2$		
-3	0.00374	0.00886	0	0.00410
-2	0.00115	0.00382	0	0.00158
-1	0.00029	0.00158	0	0.00043

Table 6. Percentage of messages decoded with different values of B_{max}

SNR_B	FastB-Cut-Decoding					FastB-4-Sets-Cut-Decoding				
	B_{max}									
	1	2	3	4	5	1	2	3	4	5
	$P_{GG} = 0.8$	$P_{BB} = 0.8$								
−3	3.49%	14.38%	24.80%	28.57%	28.76%	23.22%	18.02%	22.20%	25.42%	11.13%
−2	5.16%	22.52%	32.39%	24.98%	14.96%	27.61%	25.09%	26.56%	17.04%	3.70%
−1	8.10%	34.92%	34.77%	16.15%	6.06%	35.64%	32.21%	24.14%	7.33%	0.68%
	$P_{GG} = 0.5$	$P_{BB} = 0.5$								
−3	0.22%	9.41%	33.84%	33.83%	22.70%	7.13%	20.07%	35.07%	30.54%	7.19%
−2	0.78%	20.39%	43.13%	25.25%	10.45%	12.36%	31.98%	38.33%	15.50%	1.84%
−1	2.62%	39.57%	40.81%	13.28%	3.72%	23.95%	42.21%	28.65%	4.85%	0.34%
	$P_{GG} = 0.2$	$P_{BB} = 0.8$								
−3	0%	0.18%	6.57%	29.51%	63.75%	0.17%	2.12%	8.19%	45.62%	43.90%
−2	0%	1.56%	20.51%	41.17%	36.76%	0.68%	6.96%	25.23%	50.45%	16.68%
−1	0.04%	7.75%	40.79%	35.68%	15.74%	3.18%	19.99%	44.45%	28.76%	3.62%
	$P_{GG} = 0.8$	$P_{BB} = 0.2$								
−3	17.09%	49.87%	24.72%	6.55%	1.78%	54.50%	31.98%	11.89%	1.55%	0.09%
−2	25.58%	52.37%	17.68%	3.64%	0.73%	63.89%	29.17%	6.44%	0.48%	0.01%
−1	36.12%	51.60%	10.25%	1.68%	0.35%	73.37%	23.83%	2.68%	0.12%	0%

larger B_{max}) with FastB-4-Sets-Cut-Decoding algorithm than with FastB-Cut-Decoding algorithm. In almost all cases, the percentage of messages which decoding needed $B_{max} = 5$ with these algorithm is below 8% and for more than 60% of messages the decoding ended with $B_{max} \leq 3$. Exception of this are only the cases when $P_{GG} = 0.2, P_{BB} = 0.8, SNR_B \leq -2$ and $P_{GG} = 0.8, P_{BB} = 0.8, SNR_B = -3$, Also, for $P_{GG} = 0.8, P_{BB} = 0.2$ more than half of decodings ended with $B_{max} = 1$. This means that for these channel parameters FastB-4-Sets-Cut-Decoding algorithm is faster than Burst-4-Sets-Cut-Decoding algorithm.

5 Conclusion

In this paper a new modification of Fast-Cut-Decoding and Fast-4-Sets-Cut-Decoding algorithms for RCBQ are proposed. In order to customize these algorithms for correction of burst errors in the new algorithms, called FastB-Cut-Decoding and FastB-4-Sets-Cut-Decoding algorithms, we include an interleaver in coding algorithm and the corresponding deinterleaver in the decoding algorithm. In this way, we obtained better results for packet-error and bit-error probabilities than with the previously defined (Burst-Cut-Decoding and Burst-4-Sets-Cut-Decoding) algorithms of RCBQ for transmission through a burst channels. Also, from the presented percentages of messages which decoding ends with different values of B_{max} we can conclude that for some channel parameters FastB-Cut-Decoding and FastB-4-Sets-Cut-Decoding algorithms provide faster decoding.

Acknowledgment. This research was partially supported by Faculty of Computer Science and Engineering at "Ss Cyril and Methodius" University in Skopje.

References

1. Gligoroski, D., Markovski, S., Kocarev, L.j.: Error-correcting codes based on quasi-groups, In: Proceedings of 16th International Conference on Computer Communications and Networks, New York, NY, USA, pp. 165–172. IEEE (2007)
2. Gligoroski, D., Markovski, S., Kocarev, Lj.: Totally asynchronous stream ciphers + redundancy = cryptcoding. In: Aissi, S., Arabnia, H.R. (eds.) International Conference on Security and management, SAM 2007, pp. 446–451. CSREA Press, Las Vegas (2007)
3. Labiod, H.: Performance of Reed Solomon error-correcting codes on fading channels. In: Proceedings of the IEEE International Conference on Personal Wireless Communications (Cat. No. 99TH8366), Jaipur, India, New York, NY, USA, pp. 259–263. IEEE (1999)
4. Mathur, C.N., Narayan, K., Subbalakshmi, K.P.: High diffusion cipher: encryption and error correction in a single cryptographic primitive. In: Zhou, J., Yung, M., Bao, F. (eds.) ACNS 2006. LNCS, vol. 3989, pp. 309–324. Springer, Heidelberg (2006). https://doi.org/10.1007/11767480_21
5. Mechkaroska, D., Popovska-Mitrovikj, A., Bakeva, V.: Cryptcodes based on Quasigroups in Gaussian channel. Quasigroups Relat. Syst. **24**(2), 249–268 (2016)
6. Mechkaroska, D., Popovska-Mitrovikj, A., Bakeva, V.: New cryptcodes for burst channels. In: Ćirić, M., Droste, M., Pin, J.É. (eds.) CAI 2019. LNCS, vol. 11545, pp. 202–212. Springer, Cham (2019). https://doi.org/10.1007/978-3-030-21363-3_17
7. Popovska-Mitrovikj, A., Bakeva, V., Markovski, S.: On random error correcting codes based on quasigroups. Quasigroups Related Syst. **19**(2), 301–316 (2011)
8. Popovska-Mitrovikj, A., Bakeva, V., Mechkaroska, D.: New decoding algorithm for cryptcodes based on quasigroups for transmission through a low noise channel. In: Trajanov, D., Bakeva, V. (eds.) ICT Innovations 2017. CCIS, vol. 778, pp. 196–204. Springer, Cham (2017). https://doi.org/10.1007/978-3-319-67597-8_19
9. Popovska-Mitrovikj, A., Markovski, S., Bakeva, V.: Increasing the decoding speed of random codes based on quasigroups. In: Markovski, S., Gusev, M. (eds.) ICT Innovations 2012, Web proceedings, pp. 93–102. Springer, Heidelberg (2012)
10. Popovska-Mitrovikj, A., Markovski, S., Bakeva, V.: Performances of error-correcting codes based on quasigroups. In: Davcev, D., Gómez, J.M. (eds.) ICT Innovations 2009, pp. 377–389. Springer, Heidelberg (2010). https://doi.org/10.1007/978-3-642-10781-8_39
11. Popovska-Mitrovikj, A., Markovski, S., Bakeva, V.: Some new results for random codes based on quasigroups. In: Proceedings of the 10th Conference on Informatics and Information Technology, pp. 178–181 (2013)
12. Popovska-Mitrovikj, A., Markovski, S., Bakeva, V.: 4-Sets-Cut-Decoding algorithms for random codes based on quasigroups. Int. J. Electron. Commun. (AEU) **69**(10), 1417–1428 (2015)
13. Hwang, T., Rao, T.R.N.: Secret error-correcting codes (SECC). In: Goldwasser, S. (ed.) CRYPTO 1988. LNCS, vol. 403, pp. 540–563. Springer, New York (1990). https://doi.org/10.1007/0-387-34799-2_39
14. Zivic, N., Ruland, C.: Parallel joint channel coding and cryptography. Int. J. Electr. Electron. Eng. **4**(2), 140–144 (2010)

Cybersecurity Training Platforms Assessment

Vojdan Kjorveziroski$^{(\boxtimes)}$ ⓘ, Anastas Mishev ⓘ, and Sonja Filiposka ⓘ

Faculty of Computer Science and Engineering, Ss. Cyril and Methodius University,
Rugjer Boshkovikj 16, 1000 Skopje, North Macedonia
{vojdan.kjorveziroski,anastas.mishev,
sonja.filiposka}@finki.ukim.mk

Abstract. Hands-on experience and training related to the latest cyberthreats and best practices, augmented with real-life examples and scenarios is very important for aspiring cybersecurity specialists and IT professionals in general. However, this is not always possible either because of time, financial or technological constraints. For cybersecurity exercises to be effective they must be well prepared, the necessary equipment installed, and an appropriate level of isolation configured, preventing inter-user interference, and protecting the integrity of the platform itself. In recent years there have been numerous cybersecurity training systems developed that aim to solve these problems. They can either be used as cloud or self-hosted applications. These solutions vary in their level of sophistication and ease-of-use, but they all share a single goal, to better educate the cyber community about the most common vulnerabilities and how to overcome them. The aim of this paper is to survey and analyze popular cybersecurity training systems currently available, and to offer a taxonomy which would aid in their classification and help crystalize their possibilities and limitations, thus supporting the decision-making process.

Keywords: Cybersecurity exercises · Cybersecurity training platforms taxonomy · Training

1 Introduction

The cybersecurity workforce gap grows larger every year, reaching up to 4.07 million in 2019 [1]. There is an increasing need for trained professionals that are able to design secure systems from the ground-up, taking into account modern threats, as well as ever stricter laws whose aim is to safeguard customer information and the associated challenges in terms of their implementation [2]. The number of cyberattacks increases rapidly [3], with numerous examples of high profile cases making headline news, affecting the everyday lives of thousands of users and incurring millions in financial losses [4, 5].

For these reasons, there has been a steady increase in the number of cybersecurity initiatives, with many programming and system design courses including security-related modules in their lectures, and increasing the overall amount of cybersecurity related programs, thus raising the security awareness even higher and bridging the current workforce gap [6]. However, it is very important that any training is augmented with real-life

© Springer Nature Switzerland AG 2020
V. Dimitrova and I. Dimitrovski (Eds.): ICT Innovations 2020, CCIS 1316, pp. 174–188, 2020.
https://doi.org/10.1007/978-3-030-62098-1_15

examples and scenarios, so that the future cybersecurity professionals can get hands-on experience. It has been shown that learners' performance improves when there is an opportunity to experience first-hand the theoretical concepts [7].

Designing material with the aim to offer as much practical experience to participants as possible has proven challenging in the past. Namely, there are a lot of hurdles that are either significantly time consuming or difficult to overcome (not only in terms of complexity, but also financially), such as: 1) obtaining, installing and maintaining the underlying infrastructure where the exercises would be performed; 2) designing the exercises themselves and seamlessly integrating them both in the syllabus, as well as with the existing computer systems (e.g. assessment, training management) [8]; 3) ensuring the security and integrity of the system itself.

Despite all of these challenges, there are many examples of such systems being designed and used in the training process of new cybersecurity professionals [9–12]. Recently there has also been an increase in free or freemium [13] solutions which offer cybersecurity challenges to interested users. Entire online communities have been formed, such as the Open Web Application Security Project (OWASP) [14] whose aim is to produce articles, documentation, and open-source tools aimed at web application security. One of their most well-known contribution is the OWASP top 10 [15] list, which outlines the most common vulnerabilities in web applications today.

No matter the implementational details, all of these solutions share a common cause, the desire to raise the awareness for cybersecurity by offering hands-on experience, and in the process of doing so, stimulate adversarial thinking using an experimental approach [16]. In many cases this is reinforced by the psychological impact of the gamification aspect of the challenges and their induced competitive nature, i.e. they are structured in a game-like manner using objectives, points, and leaderboards.

The aim of this paper is to present, describe and compare existing cybersecurity training platforms and services and offer a standardized taxonomy for their classification. The rest of the paper is organized as follows: in section two we present the criteria for selecting a software for analysis, describe the evaluation methods, and offer a taxonomy which should aid in the classification. In section three, a brief overview of each solution is given, exploring its features and license model. In section four we compare and summarize their advantages and disadvantages. Finally, we conclude the paper with closing remarks and future research.

2 Methodology

The recent rise in the number of cybersecurity related solutions paved the way for new use-cases and scenarios. With the advances in virtualization technology and the ubiquitous nature of cloud computing, it is now possible to create full-fledged scenarios reminiscent of vulnerable systems found in the wild. For example, many training platforms are offering on-demand virtual machines susceptible to known attacks, that can be probed for weaknesses and exploited by participating users. This approach is in stark contrast to the more traditional web-based challenges approach. Furthermore, in order to increase the competitiveness among the participants, it is common to offer capture-the-flag style challenges, inspired by major cybersecurity events [17], where the user is tasked with obtaining a unique token (flag), thus validating the successful completion of the task at hand.

While there are numerous papers discussing the creation and implementation of such systems, or the integration of an existing open-source solution in cybersecurity courses [18], we are not aware of any work whose aim is to offer a comprehensive analysis of what is currently available.

During our research, two distinct types of systems were identified, which will be distinguished as either platforms or services. In the case of a platform, the system itself is completely extensible and additional challenges can be added with minimal effort. One such example is a system that offers its participants to define scenarios that are comprised of virtual machines, containers, or even allow the upload of the source code of a web application. In all cases either a known vulnerable software, or an unpatched version of a popular application is run, which others could then attempt to exploit. In this scenario, the system itself only acts as a middleman between the content creators and the content consumers and does not define the content itself. On the other hand, systems which themselves represent the vulnerable application that needs to be exploited, such as dedicated vulnerable training websites focusing on a finite number of scenarios, offering no extension path, are classified as services.

Taking into account the vast number of solutions currently available, mainly differentiated as being either cloud-based (hosted) or open-source (self-hosted), we have done the initial selection process based on the Alexa rank [19], for hosted services and on the number of GitHub stars [20], for self-hosted solutions. During this process, close attention was paid to include services and platforms from both camps, hosted and self-hosted, as to be able to compare the offerings between them. After manually testing each solution, and determining the advantages, disadvantages, and similarities with each other, we have identified a set of common parameters that are shared between them, discussed in detail below.

Two main types of systems are distinguished based on who is responsible for their hosting: 1) cloud-based, where the user can optionally register before using the service as-is; 2) self-hosted, where the source code can be used locally.

Regarding the product licensing, four different options have been identified: either completely free, with all features available to the end-user or, alternatively, a license fee has to be paid upfront before the user can access the challenges or download the necessary source files. A slight variation of these two options is the freemium model, where the user has access to certain introductory challenges, but then must upgrade to some form of premium membership before viewing the rest of the content. Finally, there are numerous examples of open-source solutions that also do not require any fees and can be freely downloaded and used by the end-users.

The number of available challenge types varies significantly, but the most common are: a) OWASP top 10 security risks which include SQL injection, cross-site scripting, authentication issues; b) networking challenges in the form of packet sniffing, man-in-the-middle attacks and replay attacks; c) exploitation challenges that aim to use a known vulnerability in some operating system or software package to gain unprivileged access to a remote resource; d) cryptography challenges that range from breaking simple cryptographic ciphers, to more complex ones, such as exploiting a vulnerability in a popular application.

The vast number of exercises, as well as their diversity, make the answer verification process (whether the user has submitted the correct answer) quite challenging. The most common patterns seen across the different solutions are either manual verification or automatic verification. Keeping in mind that some of the challenges can be quite complex, it is not rare to see cases where the only verification option is to manually acknowledge that the challenge has been solved by simply clicking a button or revealing the correct answer after a period of time has passed. This value then needs to be manually pasted in a verification box. In this scenario it is up to the user to refrain from abusing this option and prematurely revealing the right answer without trying to complete the challenge by themselves. The other option is automatic verification where the system can automatically mark the challenge as passed after detecting that the necessary modification has been made to some watched resource (e.g. a new super user has appeared after exploiting an SQL-injection vulnerability). Of course, this is not possible with all challenge types. Finally, as a slight variation of the automatic option where the user has to verify the answer by themselves, an alternative approach is given, where another user, usually the administrator, has to check the working environment and verify that the challenge has been successfully solved before awarding any points. Should the user have issues with any of the challenges, in most cases, a collaborative space in the form of forums or dedicated chat rooms is available. Regardless of the communication method, sharing of complete solutions is strictly forbidden, keeping in line with the competitive nature of the challenges.

Another differentiating factor between the various training systems is the way that the challenges are presented to the end-user. Some of them simply offer the necessary source file where the task is described, without presenting any environment where the users can try to solve the challenge. However, in most cases, there is a way in which a per-customer instance of the challenge can be created. This is either done by instantiating dedicated virtual machines and accessing them through a remote console, or by creating containers where the vulnerable software is sandboxed and remote access to the end user is provided. A variation of these approaches is the case where the raw virtual machine disks or container image files are provided to the end-user to be run locally. Finally, when it comes to web-based challenges, usually just a URL to the vulnerable application is provided, hiding any additional details about the hosting infrastructure from the users. Most of the sites take pride in the quality of their content and restrict the creation of new challenges either to administrators or to distinguished members who have accumulated a large amount of points through frequent usage. Others may be more focused on the community character and in this case either all registered members can create new content or the privilege for content moderation can be acquired for a fixed fee usually in the form of a donation or a recurring subscription. Since most of the solutions offer a leaderboard where the members are ranked, as well as an option to track the progress through the various challenges, users are usually required to register an account upon their first visit.

This can either be done through a traditional email login or by using some of the widely popular social login options.

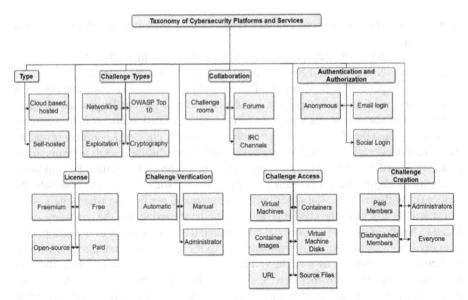

Fig. 1. Cybersecurity software taxonomy

These observations were used as a starting point in the creation of the cybersecurity software taxonomy presented in Fig. 1. The taxonomy consists of 8 categories with a total of 29 parameters, describing the nature of the solution, challenge type, licensing model, collaboration features, verification of the provided answers, authentication and authorization options, technology being used for the challenge environment as well as who can curate the content.

Using the popularity criteria discussed previously, we have narrowed down the initial selection of platforms and services to 13. In the section that follows, a brief overview of the selected solutions is provided, centered around the proposed taxonomy.

3 Solution Overview

The overview of the selected solutions has been divided into two general groups, cloud and self-hosted.

3.1 Cloud Solutions

We classify as cloud solutions all solutions that have a publicly accessible website where the user can list, preview, and attempt the challenges. In some cases, the way in which the challenges are accessed is mixed - there are both hosted challenges and challenges for which the user must download an additional virtual image locally. Nevertheless, as long as the service has a web presence around which a community is built it is classified as a cloud service.

Enigma Group Challenges [21]. A well-known web site offering over 300 challenges with over 61000 members. The challenges are divided into seven categories:

- Basic challenges where the task is clearly stated, and it is up to the user to exploit the given vulnerability. Usually the basic challenges are centered around the OWASP top 10 and automatic answer verification is provided.
- Realistic challenges, where a real-world scenario is described, and it is up to the user to decide how to tackle it. These challenges are usually more involved since the user might have to install additional software, such as port scanning tools.
- Cryptographic challenges focus on encryption, decryption, brute forcing.
- Cracking challenges where the user is provided with only a binary file and is tasked with figuring out its purpose and how to exploit it.
- Auditing challenges where a block of source code is given and based on the detected bugs in it, the user should take appropriate action to exploit the vulnerability, for a purpose which has been stated beforehand.
- Patching challenges, similar to the auditing challenges, but now the user is tasked with fixing the bugs so that the application is no longer vulnerable.
- Programming challenges where a problem is stated, and the user must write a program solving it in a very short period of time, before a timer expires.
- Steganography challenges, where files in which hidden messages are present are downloaded by the user.

One of the main advantages of Enigma group is that automatic answer verification is provided. Depending on the complexity of the challenge the user is awarded points which count towards their rank. A lot of the challenges are restricted to paying members only. A public forum, an IRC channel and personal member blogs are available, all of which can be used for discussing challenges and sharing any cybersecurity related topics. Challenge creation is exclusively done by the administrators as to ensure their quality.

Hack Yourself First [22]. Hack Yourself First is different than most of the other services discussed in the sense that it is not a full-fledged platform where various challenges can be posted, graded and points awarded. Instead, it is simply a vulnerable website with over 50 security holes that the visitor can try and exploit, such as add new entries, impersonate other users, delete data, steal cookies. There are neither points awarded for successfully exploiting a vulnerability, nor any sort of validation. The user knows that the attempt was successful if the state of the website has been altered. The site offers account registration, but this is part of the scenario and it is up to the user to exploit the registration process. No official collaboration spaces are provided, but a video course [23] is available discussing the main concepts behind the vulnerabilities present on the site and reinforcing the message of how bad coding practices can negatively impact users. The underlying database is periodically rebuilt so that new users can start fresh.

Hack This Site [24]. Hack This Site is similar to the previously introduced Enigma group challenges, since they both offer various tasks which are divided into subgroups. In the case of Hack This Site the term mission is used to identify a challenge, and the mission groups are basic, realistic, application, programming, phone phreaking, JavaScript, forensic and steganography missions.

Most of the groups closely match the ones described in the Enigma group section, however, one of them is exclusive to Hack This Site - the phone phreaking missions. Here the users are tasked with the exploitation of a public branch exchange (PBX) telephone system, where they must use a real phone to conduct the mission.

The site requires the users to be registered before they can access any of its content and has a strong community character with regular blog posts, articles, and open forums where the users can discuss cybersecurity related topics. Most of the missions support automatic verification either through detection that the required actions have been performed or through capture the flag style answers, where the user has to provide the answer in a textbox before points are awarded. All of the challenges are free to access once the user has created an account; the site is funded by donations and through the online shop [25] where branded apparel as well as everyday home and office items are sold.

Root-Me [26]. Root-Me is a commercial platform that offers some of its content for free, while the rest is restricted behind a subscription paywall. Premium accounts can either be obtained by paying a monthly or yearly fee, or by paying a much smaller fee for a contributor's account and then creating new challenges on the platform. Challenge verification is done by submitting a flag that has been obtained and points are awarded for a successful competition of a challenge. Scoreboards and public rankings are available to encourage the competitiveness between the users. Each challenge has a hall-of-fame section where all the users that have previously completed the given challenge are listed.

Most of the challenges are hosted in virtual machines, where access is restricted to the IP addresses of the currently online members on the platform. Less experienced users have the option of taking a learning path which groups the available challenges and orders them based on difficulty and required prerequisite knowledge. In this way, novice users can first solve challenges related to programming basics, before moving on to networking concepts, and finally exploitation of web applications or other systems. Forums and an IRC channel are available so that users can seek help and discuss security related topics.

Try Hack Me [27]. Try Hack Me is one of the more advanced platforms that were evaluated. The challenges are divided into rooms. Each room is represented by a challenge description in a blog-post format where the vulnerability is described, and the necessary tools are listed. The users progress through a challenge by capturing flags during the exploitation process and validating them on the room's home page. Most of the rooms have virtual environments associated with them, where the vulnerable software is installed in a virtual machine accessible only to the user attempting the challenge. Access to the virtual machines is provided through a virtual private network (VPN) connection. Using this architecture, it is possible to host any type of challenge, from exploitation of a vulnerability in an operating system's kernel to simpler web-based challenges.

The site offers a paid tier whose benefits are faster setup of the virtual environment and access to a wider selection of rooms. An additional benefit awarded to paying customers is a dedicated Kali Linux [28] virtual machine that users can access remotely. Login is only required for VM creation and answer validation, while the content of the free rooms can be accessed even without an account.

Any member can submit a new challenge, along with an associated virtual environment. It is up to the user to design the whole scenario and submit a virtual disk image to the platform, which would later be used as a master image for the environment creation. Also, many of the rooms freely offer the base image to the end-users, so that they can run the scenario locally instead of using the cloud resources of the platform. Any room can be forked, making a 1:1 copy of its content which can be improved by changing the scenario or modifying the base image. Each room has a dedicated forum associated with it, along with a writeups section where other users can explain the process of solving the presented challenges in detail. A scoreboard is used for listing the top performing users in each of the challenges.

Should the user want to learn a new skill, for example "web fundamentals", appropriate learning paths are provided, where through a series of exercises the fundamental concepts of a given topic are presented.

Hack Me [29]. Hack Me is similar to Try Hack Me in the sense that it allows any member to create full-fledged scenarios. However, unlike Try Hack Me, these scenarios are not sandboxed in a remote virtual environment and are limited to uploading PHP scripts. Additional required PHP libraries can be selected during the creation process as well as an optional MySQL instance, along with seed data. Once the challenge has been uploaded and made public, users can attempt to solve it. For each user, a unique sandbox of the scenario is created, accessible via a public random URL.

No native way in which answers can be validated is offered, it is up to the challenge developer to include verification logic in their scripts. A comment section hosted by a third-party service is available for each challenge. No subscription is required, except for a creation of a free account before the challenges can be accessed.

3.2 Self-hosted Solutions

Self-hosted solutions differ from their cloud counterparts in the sense that they offer no infrastructure of their own, instead only the source code is provided to the user. This is different from the option where a cloud solution would offer either the base virtual image for download or the source files of a vulnerable web application, since in the hosted case there is a dedicated location for challenge download, collaboration with other members and answer submission.

Even though the self-hosted experience might seem barebones and lacking at first, it offers greater flexibility since the user can inspect, modify, and adapt the code both of the software itself and any included challenges as they see fit. Additionally, this approach offers the option of altering the software so that it can be integrated with any existing systems in a given corporation or institution, which can prove useful in terms of users import from an external database or export of any points to another system.

Of course, the main drawback is that users first need to have access to an infrastructure where they can host the applications, and also the necessary system administration experience to deploy them. This is not always a realistic expectation, especially in an academic environment where students might have different levels of system expertise. However, this can be overcome by either deploying a central instance internally, in the case where the application supports multiple users and can be used concurrently without any interference, or multiple instances which can be shared among teams of users.

OWASP Juice Shop [30], OWASP NodeGoat [31], OWASP Mutillidae II [32], OWASP WebGoat [33]. All of the OWASP Juice Shop, OWASP NodeGoat, OWASP Mutillidae II and OWASP WebGoat applications focus on presenting and educating the user about the OWASP top 10 vulnerabilities. The main difference between them is the technology stack with which they have been developed, allowing the user to choose the technology that they are most familiar with. This is a major advantage, having in mind that all of the applications are completely open-source and the user can either choose to treat the system as a black-box and discover and exploit the vulnerabilities exclusively through the web interface, or take a look at the source code and explore all of the bad code practices that have been employed, finding weaknesses along the way.

Many of these projects have also been extended by the community to offer additional modules and integrations. One such example is the OWASP Juice Shop command line tool [34] that can generate capture the flag style questions and answers for popular capture the flag contest hosting platforms like Facebook Capture the Flag and CTFd. This approach makes it possible to set up a local contest or to include these tools as part of a course with minimal effort required.

The installation procedure varies between the projects, but since in all cases the source code is readily available, it usually comes down to installing a database and the appropriate application server. Most of them also offer either dedicated installation scripts, virtual machine images, or Docker images as to ease the installation process.

Reinforcing the educational purpose of these applications, some of them have published instructional material that explains the vulnerabilities in detail, both from a theoretical point of view, as well as from a practical one. This material is available either as an ebook, as is the case with OWASP Juice Shop [35], or as sections within the application itself, like OWASP WebGoat.

Haaukins [36]. Haaukins uses YAML files and either Docker images or VirtualBox OVA files for the instantiation of new challenges. The YAML file specifies all of the container images or VirtualBox templates from which the scenario should be created, along with additional parameters, such as memory limits and environment variables. All of the containers and VMs that are part of a given challenge are isolated in a separate network. Users access a Kali Linux instance deployed inside this network through a web RDP client. As was the case with other applications, Haaukins can also integrate with the popular CTFd platform for answer verification.

Unfortunately, the platform itself does not offer more advanced challenge customization through the YAML file itself, it is up to the image developer to integrate it into the base image. However, this simplifies the deployment since no additional orchestration tool is needed.

Facebook Capture the Flag [37]. Facebook Capture the Flag is a software for hosting capture the flag style contests using a video game-like interface. All of the questions are associated with a real-world country placed on a geographic map and it is up to the users who are divided into teams to try and capture as many territories as possible by correctly answering the accompanying questions. The software itself does not come with any prepopulated question bank, it is up to the administrators to develop questions that best reflect their type of event and usage scenario. One of the main benefits of the software is the possibility to import and export questions in popular markup formats, thus easing the integration process with external systems, as was the case with the aforementioned OWASP Juice Shop application, where an additional module is capable of generating questions regarding all of the present vulnerabilities.

Facebook Capture the Flag supports multiple ways in which user accounts can be created and authenticated, such as email based registration, restricted registration based on pre-generated invite codes, or integration with external systems such as the Lightweight Directory Access Protocol (LDAP). The challenge parameters are very granular and can be tweaked by the administrators, allowing the challenges to be time restricted, alter the number of points awarded depending on the complexity of the question, whether a given country can be conquered more than once (whether a given question can be answered multiple times), etc.

CTFd [38]. Another application for organizing and hosting capture the flag challenges, similar to Facebook Capture the Flag. However, CTFd supports more advanced question types such as unlockable challenges, where the user is required to correctly answer a given question before being allowed to progress to the next one; multiple choice challenges, where instead of providing the answer in a text field, the user can choose the correct answer from a list of given answers; dynamic challenges where each subsequent correct answer to a question lowers the amount of points awarded; manual verification challenges where a privileged user is required to grade the question and finally programming challenges. Programming challenges are the most advanced challenge type since the user is tasked with writing a program which needs to solve the described problem. Additionally, the platform is able to evaluate the code locally by supporting some of the most popular languages such as Java, Python, C/C ++ and NodeJS. Once submitted, it is evaluated whether a correct solution has been provided for the given standard input (stdin) by matching the standard output (stdout) of the program with the expected stdout provided by the administrator during the challenge creation.

Similar to other offerings, users can also be grouped into teams and a dedicated statistics page is available providing useful information about the registered users and challenges.

4 Solution Comparison

Table 1 compares all of the applications that were discussed in the previous section using the taxonomy presented in Fig. 1. To increase readability, some of the parameters have been abbreviated, for which a detailed explanation is given in the footer of the table. A note should also be made about the automatic verification parameter when it comes to self-hosted applications. Even though many of them do not support automatic answer verification, a tick is present in the corresponding column if there is a possibility to integrate the application with an external system which can provide the answer checking functionality, for example a capture the flag hosting application. An additional column, not included in the taxonomy is the rank column, stating the relative popularity of the given service, where numbers prefixed with an A, in the case of cloud platforms, represent the Alexa rank of the service, while the ones prefixed with G the number of stars on the GitHub code hosting platform.

Based on the results presented in Table 1, according to their Alexa rank, cloud-based solutions where only static challenges are provided are less popular. This is due to the fact that they do not offer dedicated virtual machines or containers which users can access, unlike their counterparts which do offer their users full-fledged environments. It can be argued that challenge creation also plays a role, with community-based platforms such as Root-Me, Try Hack Me and Hack Me being more popular than their counterparts such as, Enigma Group Challenges and Hack Yourself First, where challenges are only created by the site administrators.

When it comes to the self-hosted solutions, Facebook CTF is a particularly popular option, whose popularity can be attributed to its interface design and the fact that it gives an additional incentive to the user with its world domination aspect. CTFd is an option that has either been integrated as an optional plugin or as a dependency to some of the other platforms, such as Haaukins. The applications under the OWASP umbrella also enjoy high popularity and very active development. Their main advantage is the fact that they are not merely vulnerable applications where the user has to possess prior knowledge to find and exploit the vulnerabilities, but also offer dedicated sections explaining bad coding practices and how they can be overcome. Some of the options, like Mutillidae II even offer a secure and non-secure mode of operation, along with web-based reset controls which can bring the application to its initial state after any database modifications.

While cloud-based solutions are easier to use and do not require extensive maintenance, almost all require payment to unlock additional features. Additionally, cloud security training platforms might be attractive targets for potential hackers which may lead to information leakage. To overcome these problems, free and open-source solutions need to be chosen with permissive licenses that allow altering of the source-code so that they can be integrated with existing systems.

In summary, solutions that offer a more realistic approach to the challenges, with virtual machines or containers are more popular and offer a better learning experience for the end-user. However, these options are also much more complex, dealing with issues such as provisioning, network isolation and remote access.

Table 1. Software comparison

		Enigma	HYF	HTS	Root-Me	TryHackMe	HackMe	OWASP JS	OWASP NG	OWASP ML	OWASP WG	Haaukins	Facebook CF	CTFd
Type		cb	cb	cb	cb	cb	cb	sh	sh	sh	sh	sh	sh	sh
License		fm	fr	fr	fm	fm	fr	os	os	os	os	os	os	os
C. Types	n	✗	✗	✗	✓	✓	✓	✗	✗	✗	✗	✓	✗	✗
	o	✓	✓	✓	✓	✓	✓	✓	✓	✓	✓	✓	✗	✗
	e	✓	✗	✓	✓	✓	✓	✗	✗	✗	✗	✓	✗	✗
	c	✓	✗	✗	✓	✓	✓	✗	✗	✗	✗	✓	✗	✗
C. Ver.	au	✓	✗	✓	✓	✓	✓	✓	✗	✗	✓	✓	✓	✓
	m	✗	✓	✗	✗	✗	✓	✗	✓	✓	✓	✓	✗	✗
	ad	✗	✗	✗	✗	✗	✗	✗	✗	✗	✗	✗	✗	✓
Collab.	cr	✗	✗	✗	✗	✓	✓	✗	✗	✗	✗	✗	✗	✗
	f	✓	✗	✓	✓	✓	✗	✗	✗	✗	✗	✗	✗	✗
	ic	✓	✗	✓	✓	✗	✗	✗	✗	✗	✗	✗	✗	✗
C. Access	vm	✗	✗	✗	✓	✓	✗	✗	✗	✗	✗	✓	✗	✗
	c	✗	✗	✗	✗	✗	✗	✗	✗	✗	✗	✓	✗	✗
	ci	✗	✗	✗	✗	✗	✗	✗	✗	✗	✗	✓	✗	✗
	vd	✗	✗	✗	✗	✓	✗	✗	✗	✗	✗	✓	✗	✗
	u	✓	✓	✓	✓	✓	✓	✓	✓	✓	✓	✓	✓	✓
	sf	✓	✗	✓	✓	✓	✓	✓	✓	✓	✓	✓	✗	✓
AA	a	✗	✓	✗	✗	✗	✓	✓	✗	✗	✓	✗	✗	✗
	e	✓	✗	✓	✓	✓	✓	✓	✓	✓	✗	✓	✓	✓
	s	✗	✗	✗	✗	✗	✗	✓	✗	✗	✗	✗	✓	✗
CC	pm	✗	✗	✗	✓	✓	✗	✗	✗	✗	✗	✗	✗	✗
	a	✓	✓	✓	✓	✓	✓	✗	✗	✗	✗	✓	✓	✓
	dm	✗	✗	✗	✓	✗	✗	✗	✗	✗	✗	✗	✗	✗
	e	✗	✗	✗	✗	✓	✓	✗	✗	✗	✗	✗	✗	✗
Rank		A. 3.3M	A. 2.4M	A. 93K	A. 91K	A. 533K	A. 734K	G. 2.9K	G. 1.1K	G. 330	G. 3.1K	G. 71	G. 6.2K	G. 2.5K

Type - cb (cloud-based) or sh (self-hosted); **License** - fm (freemium), fr (free), os (open-source), p (paid); **C. Types:** Challenge Types - n (networking), o (OWASP top 10), e (exploitation), c (cryptography); **C. Ver.:** Challenge Verification - au (automatic), m (manual), ad (administrator); **Collab:** Collaboration - cr (challenge rooms), f (forums), ic (irc chat); **C. Access:** Challenge Access - vm (virtual machine), c (container), ci (container images), vd (virtual machine disks), u (url), sf (source files), **AA:** Authentication and Authorization - a (anonymous), e (email login), s (social login); **CC:** Challenge Creation - pm (paid members), a (administrators) , dm (distinguished members), e (everyone); **Rank** - A (Alexa), G (GitHub stars).

5 Closing Remarks

The increasing cybersecurity requirements of modern applications and the popularization of this topic through numerous high-profile breaches that have made newspaper headlines around the world, have led to the proliferation of training platforms and services that aim to provide challenging tasks as to enhance the practical cybersecurity experience globally. However, faced with so many options, users, as well as system integrators who would like to use some of these systems in their existing infrastructure, are faced with difficult choices, since no comparative analysis is available. In order to mitigate this, a survey of publicly available solutions was made and through the analysis of their characteristics and main features, a taxonomy was proposed which can aid in cybersecurity training software classification.

The proposed cybersecurity taxonomy includes parameters such as deployment type, license model, types of challenges supported, methods for verifying the challenge solutions, challenge deployment options, user authentication and authorization options and finally challenge creation. An overview of the two main application categories, cloud and self-hosted was given. A short description of 13 platforms and services, selected based on their popularity, followed by a more thorough discussion about the various advantages and disadvantages stemming from their chosen architecture was provided. Taken together, these resources should provide a more complete picture about the currently available cybersecurity training solutions, and the various use-cases supported.

Future work will focus on the design and implementation of a general-purpose cybersecurity training platform to be used as part of cybersecurity courses, which will be based on the best-practices and some of the open-source tools introduced.

References

1. (ISC)^2: 2019 Cybersecurity Workforce Study (2019). https://www.isc2.org/-/media/ISC2/Research/2019-Cybersecurity-Workforce-Study/ISC2-Cybersecurity-Workforce-Study-2019.ashx. Accessed 26 Feb 2020
2. Poritskiy, N., Oliveira, F., Almeida, F.: The benefits and challenges of general data protection regulation for the information technology sector. DPRG (2019). https://doi.org/10.1108/DPRG-05-2019-0039
3. Department for Digital, Culture, Media & Sport: Cyber Security Breaches Survey 2019. https://assets.publishing.service.gov.uk/government/uploads/system/uploads/attachment_data/file/813599/Cyber_Security_Breaches_Survey_2019_-_Main_Report.pdf. Accessed 25 Feb 2020
4. Ghafur, S., Kristensen, S., Honeyford, K., Martin, G., Darzi, A., Aylin, P.: A retrospective impact analysis of the WannaCry cyberattack on the NHS. NPJ Digital Med. (2019). https://doi.org/10.1038/s41746-019-0161-6
5. Berghel, H.: Equifax and the latest round of identity theft roulette. Computer (2017). https://doi.org/10.1109/MC.2017.4451227
6. Lopez-Cobo, M., et al.: Academic offer and demand for advanced profiles in the EU. Artificial Intelligence, High Performance Computing and Cybersecurity, JRC113966. Joint Research Centre (Seville site) (2019). http://publications.jrc.ec.europa.eu/repository/handle/JRC113966

7. Bell, R.S., Sayre, E.C., Vasserman, E.Y.: A Longitudinal study of students in an introductory cybersecurity course. In: 2014 ASEE Annual Conference & Exposition. ASEE Conferences, Indianapolis, Indiana (2014)

8. Shumba, R.: Towards a more effective way of teaching a cybersecurity basics course. SIGCSE Bull. (2004). https://doi.org/10.1145/1041624.1041671

9. Furfaro, A., Piccolo, A., Parise, A., Argento, L., Saccà, D.: A cloud-based platform for the emulation of complex cybersecurity scenarios. Future Gener. Comput. Syst. (2018). https://doi.org/10.1016/j.future.2018.07.025

10. Acosta, J.C., McKee, J., Fielder, A., Salamah, S.: A platform for evaluator-centric cybersecurity training and data acquisition. In: MILCOM 2017 - 2017 IEEE Military Communications Conference (MILCOM). 2017 IEEE Military Communications Conference (MILCOM), Baltimore, MD, 23–25 October 2017, pp. 394–399. IEEE. https://doi.org/10.1109/MILCOM.2017.8170768

11. Kalyanam, R., Yang, B.: Try-CybSI: an extensible cybersecurity learning and demonstration platform. In: Zilora, S., Ayers, T., Bogaard, D. (eds.) Proceedings of the 18th Annual Conference on Information Technology Education - SIGITE 2017. the 18th Annual Conference, Rochester, New York, USA, pp. 41–46. ACM Press, New York (2017). https://doi.org/10.1145/3125659.3125683

12. Mirkovic, J., Benzel, T.: Teaching cybersecurity with DeterLab. IEEE Secur. Privacy 10(1), 73–76 (2012). https://doi.org/10.1109/MSP.2012.23

13. Kim, W.: A practical guide for understanding online business models. Int. J Web Inf. Syst. (2019). https://doi.org/10.1108/IJWIS-07-2018-0060

14. OWASP Foundation, the Open Source Foundation for Application Security. https://owasp.org/. Accessed 27 Feb 2020

15. OWASP Top 10. https://owasp.org/www-project-top-ten/. Accessed 19 Feb 2020

16. Schneider, F.B.: Cybersecurity education in universities. IEEE Secur. Privacy 11(4), 3–4 (2013). https://doi.org/10.1109/MSP.2013.84

17. Nunes, E., Kulkarni, N., Shakarian, P., Ruef, A., Little, J.: Cyber-deception and attribution in capture-the-flag exercises. In: Jajodia, S., Subrahmanian, V.S.S., Swarup, V., Wang, C. (eds.) Cyber Deception, pp. 151–167. Springer, Cham (2016). https://doi.org/10.1007/978-3-319-32699-3_7

18. Chicone, R., Burton, T.M., Huston, J.A.: Using Facebook's open source capture the flag platform as a hands-on learning and assessment tool for cybersecurity education. Int. J. Concept. Struct. Smart Appl. 6(1), 18–32 (2018). https://doi.org/10.4018/IJCSSA.2018010102

19. Alexa. Keyword Research, Competitive Analysis and Website Ranking. https://www.alexa.com/. Accessed 25 Feb 2020

20. GitHub Stars. https://help.github.com/en/enterprise/2.13/user/articles/about-stars. Accessed 27 Feb 2020

21. Enigma Group Challenges. Web application security training. https://www.enigmagroup.org/. Accessed 18 Feb 2020

22. Hack Yourself First. https://hack-yourself-first.com/. Accessed 18 Feb 2020

23. Hunt, T.: Hack Yourself First: How to go on the Cyber-Offense. https://app.pluralsight.com/library/courses/hack-yourself-first/table-of-contents. Accessed 19 Feb 2020

24. Hack This Site. https://www.hackthissite.org/. Accessed 18 Feb 2020

25. Hack This Site Online Shop. https://www.cafepress.com/htsstore. Accessed 25 Feb 2020

26. Root Me. https://www.root-me.org/. Accessed 18 Feb 2020

27. Try Hack Me. https://tryhackme.com/. Accessed 18 Feb 2020

28. Kali Linux. Penetration Testing and Ethical Hacking Linux Distribution. https://www.kali.org/. Accessed 25 Feb 2020

29. Hack Me. https://hack.me/. Accessed 18 Feb 2020

30. OWASP Juice Shop. https://owasp.org/www-project-juice-shop/. Accessed 18 Feb 2020
31. OWASP NodeGoat. https://owasp.org/www-project-node.js-goat/. Accessed 18 Feb 2020
32. OWASP Mutillidae II. https://github.com/webpwnized/mutillidae. Accessed 18 Feb 2020
33. OWASP WebGoat. https://owasp.org/www-project-webgoat/. Accessed 18 Feb 2020
34. OWASP Juice Shop CTF CLI. https://www.npmjs.com/package/juice-shop-ctf-cli. Accessed 25 Feb 2020
35. Kimminich, B.: Pwning OWASP Juice Shop (2019). https://bkimminich.gitbooks.io/pwning-owasp-juice-shop/content/
36. Haaukins. A Highly Accessible and Automated Virtualization Platform for Security Education. https://github.com/aau-network-security/haaukins. Accessed 18 Feb 2020
37. FBCTF. Platform to host Capture the Flag competitions. https://github.com/facebook/fbctf. Accessed 18 Feb 2020
38. CTFd. https://github.com/CTFd/CTFd. Accessed 18 Feb 2020

Real-Time Monitoring and Assessing Open Government Data: A Case Study of the Western Balkan Countries

Vigan Raça[1], Nataša Veljković[2(✉)], Goran Velinov[1(✉)], Leonid Stoimenov[2(✉)], and Margita Kon-Popovska[1(✉)]

[1] Faculty of Computer Sciences and Engineering, Ss. Cyril and Methodius University, St. Rugjer Boshkovikj 16, Skopje, North Macedonia
{vigan.raca,goran.velinov,margita.kon-popovska}@finki.ukim.mk
[2] Faculty of Electronic Engineering, University of Nis, St. Aleksandra Medvedeva 14, 18106 Niš, Serbia
{natasa.veljkovic,leonid.stojmenov}@elfak.ni.ac.rs

Abstract. This paper focuses on real-time monitoring of open data in the Western Balkan countries and assessment of dataset quality using five star methodology introduced by Berners-Lee. We have designed a framework and an application prototype that uses existing Application Programming Interfaces (API) of open data portals to collect information in real-time and conduct assessment on dataset quality for the Balkan countries. The obtained data has been evaluated and assessment results are visualized and presented in a dashboard. The results show that Balkan countries have 2.5/5 level on data quality on average.

Keywords: Open government data · Datasets · Organizations · Western Balkans · Data quality · Assessment

1 Introduction

The concept of 'open data' has been around for at least 13 years [1]. Though its definition differs depending on the literature used, most of the concept definitions are built on the 'right to information' as a public right [2]. The last two decades have seen a lot of efforts - and rhetoric - about the direction of the public service and the modern state in the context of the latest digital era. Two main streams of discussions involve the idea of open government data (OGD) and the quality issues of open datasets published by public sector bodies. However, the two are more connected than it appears, because being open implies removing barriers to ensure re-use by anyone and it is becoming more obvious that the quality of OGD is affecting its reuse [3]. However, reusing open data is not a straightforward exercise, but it is the only way to effectively use such data [4, 5]. This research highlights basic points related to working with open data, which are based on experience over cases of typical public sector datasets published with the intent to be reused. Despite the fact that there are considerable research efforts addressing certain

© Springer Nature Switzerland AG 2020
V. Dimitrova and I. Dimitrovski (Eds.): ICT Innovations 2020, CCIS 1316, pp. 189–201, 2020.
https://doi.org/10.1007/978-3-030-62098-1_16

specific areas related to producing good quality open data, relatively less attention is paid to what challenges consumers of open government data face.

The objective of this research was to identify available datasets published by public institutions in the Western Balkan countries (WB) with the aim of measuring the quality of datasets. The process starts by collecting information on the actual number of organizations publishing datasets in the following countries: Albania, North Macedonia, Montenegro, Kosovo, Bosnia and Herzegovina (BiH) and Serbia. After identifying these organizations/public agencies, we started collecting all available datasets, extracting dataset formats and finally assessing dataset quality based on the 5 star model defined by Berners Lee [6]. The methodology is based on a case study approach augmented with some data quality theoretical aspects. An application prototype is designed for real-time monitoring and assessing dataset quality. Within the prototype a web-service is developed with the aim of using the APIs of OGD portals for collecting information on: datasets, public member organizations and dataset formats. Through the APIs we were able to collect open data in real-time. A web portal is created, as an entry point but also to serve as a dashboard for visualization and presentation of the results.

In this paper we also considered a typology of open data quality issues. We investigated how open are WB countries' governments in terms of the number of datasets published and public organizations involved in open data publishing. This research also draws on the distinctions among WB countries in the context of open data publishing format and specifics in terms of their quality, if any, for each country.

The rest of the paper is organized as follows. Second section reviews open data with focus on open data portals used for data publishing. In the third section we discuss the existing approaches to measure open data quality in general as well approaches based on the 5 star model and a brief comparison with our approach including advantages and expected results. The fourth section, discusses the research questions and presents the methodology used in this research. The fifth section, discusses the results and provides a comparison among the countries. At the end, this research provides conclusions and recommendations for future research related to open dataset quality dimension.

2   Open Data

Initially the idea of open data was brought in the Memorandum of Transparency and Open Government signed by Obama in 2009 [7], aimed at increasing transparency and accountability of the government. In 2011, a new initiative named Open Government Partnership (OGP) was established [8]. This initiative gathers seventy-eight countries and twenty local members averaging about two billion people. All WB countries, apart from Kosovo, are members of OGP.

The concept of open data means that data is accessible to anyone and published in a machine-readable format, which can be freely used, re-used, and redistributed by anyone with no cost. Open government data is a narrower concept compared to open data and it refers to data and information produced or commissioned by the government or government controlled agencies' [9].

The OGD are a collection of information stored in the form of datasets. OGD are considered public data and non-sensitive data, they include data related to the environment, traffic, infrastructure, planning, transport, education, health, crime etc. [10].

The use of information technology for collecting, processing and analyzing information enabled huge volumes of data to be stored on daily basis on government web portals, the so called Open Data Portal (ODP), which became a key entry point for open data. ODP hosts a collection of datasets and provides browser, search and filter functionalities over these datasets. Usually, ODPs offer an API, which allows the publisher on one hand to upload and update data and the user on the other hand to retrieve resources automatically. The most common ODP frameworks include CKAN, DKAN, OpenDataSoft, Junar, ArcGis Open Data and SOCRATA [11].

3 Data Quality Frameworks and Considerations

Researchers and scientists have promoted the concept of data quality, but it is also recognized by the private sector, which revealed its commercial potential. Data can be defined in various formats, and the scientific literature discusses and provides different concepts about data quality. Years ago the data quality term have been used for providing better results including (statistical data, textual data, visual data etc.), but later the aim of data quality was changed. In the context of public institutions, high data quality translates into good service and good relationships between governments and citizens [12]. Dawes and Helbig consider data quality as "one of the main conditions of successful reuse" [13].

Furthermore, there are specific regulations controlling the release of data. Thus, exporting controlled data to OGD portal presents some level of legitimacy on the part of the data publisher. This setting has an impact on the way of producing high quality data as well as the way how open data is generated from the raw data. Transforming raw data into open datasets is usually done through forms defined by the corresponding laws for data protection.

Despite the difference in context, the open data quality frameworks became very important. One open data quality dimension considers technical standards and abilities, processes and outcomes of producing and managing datasets, time relevance of data (i.e. if out-of-date) and/or if dataset is frequently updated. Another important element of data quality is the availability and accessibility of various types of data or data in different categories [14].

It is important to note that in order to access quality data one should not only inspect the content of the dataset, but also usability by end users [15]. Another open data quality measuring characteristic has been demonstrated in the work of Zaveri et al. who has defined and operationalized 68 metrics among 6 dimensions [16, 17]. In addition, there exist several results of frameworks dealing with a wider range of quality dimensions – but these approaches are mainly focused on the assessment of datasets [18, 19].

This research addresses quality of open government datasets in the context of dataset publication format. In comparison to other approaches that have used additional dimensions for dataset quality assessment, we employ only the 5 star model. Moreover, different from other methodologies, this research performs dataset quality assessment processing in real time by showing results in a dashboard. Also there is a possibility for further analysis in different periods of time in order to show if countries are improving dataset quality through time.

4 Methodology

The methodology consists of several steps that lead the final results - that of ranking OGD quality in WB countries..

Step1: Analysis of national OGD portals for six WB countries (Albania, North Macedonia, Montenegro, Kosovo, BiH and Serbia). This implies investigating opportunities that OGD portals offer including APIs for accessing data, types of APIs, documentation of APIs and most importantly dataset structure (i.e. the metadata describing datasets).

Step 2: For real-time monitoring and assessing of open data quality, an application prototype is designed and it consists a web service that uses APIs for searching and collecting of datasets from OGD portals. The aim of this component is fast and real time monitoring, assessment and processing of available datasets.

Step 3: The backend component is designed to analyze incoming data. The analysis is based on the 5 star model, which identifies and classifies datasets based on data formats using a classification metric ranging from 1 to 5 stars, as shown in Table 1. Using this model, we interpret the quality of datasets for each WB country.

Step 4: A frontend dashboard component will be used as real-time information stream on open data quality assessment.

Table 1. Five star model for dataset assessment

Rates	Description	Key
★	*on the web (whatever format) pdf, image, doc*	OL
★★	*machine-readable structured format .xsl, .xlsx*	RE
★★★	*non-proprietary format, csv*	OF
★★★★	*RDF Standards*	URI
★★★★★	*Linked RDF*	LD

5 Analysis of OGD Portals in Western Balkan

In WB, the open government partnership is considered the main promoter of data opening and establishing open data portals. Table 2 shows OGD portals (URLs) in WB countries used in our research.

Most of WB countries base their open data portals on existing open source frameworks, more precisely Comprehensive Knowledge Archive Network (CKAN) [20] apart from BiH which uses Drupal based Knowledge Archive Network (DKAN) [21] and Albania which has developed its own API but is based on DKAN, so we have classified it as DKAN's family.

CKAN is an open-source data platform developed by the Open Knowledge Foundation, a non-profit organization that aims to promote the openness of all forms of knowledge. In other words, CKAN is a tool for making open data portals. It helps to

Table 2. OGD Portals for West Balkan countries

Country	URL	Portal
Kosovo	https://opendata.rks-gov.net/	CKAN
North Macedonia	https://otvorenipodatoci.gov.mk/	CKAN
Serbia	https://data.gov.rs	CKAN
Bosna and Herzegovina	https://opendata.ba	DKAN
Albania	http://opendata.gov.al/	DKAN
Montenegro	https://data.gov.me/	CKAN

manage and publish data collection. Thus, it is used by national and local governments, research institutions, and other organizations that collect large amounts of data [20].

DKAN is an open data platform, based on the CKAN, with a full suite of cataloging, publishing and visualization features that allows governments, nonprofits and universities to easily publish data to the public [21]. The OGD Portals provide the API not only to consume their datasets, but for extracting datasets by individual institutions as well. It means that member organizations use the API for extracting their data in a proper way. Based on that, the API has a bidirectional function, to collect datasets (get) from web portal and to extract data (post) to the web portal. In our application prototype we use only the first function.

An application program interface (API) is a set of routines, protocols, and tools for building software applications. Basically, an API specifies how software components should interact with each other [22].

API for utilizing open datasets and other relevant information is present in all six OGD Portals; however, the type of API differs. Both CKAN and DKAN have API documentation, which can be analyzed to form proper calls in order to get the data. Even so, Serbia and Montenegro developed their own API based on CKAN and they also provide technical documentation on its usage. It is important to note that for some OGD portals (Kosovo and North Macedonia) even they use CKAN it is not available for data consume. For those cases, we have designed a JavaScript, which reads information directly from front-end of portals but not through API.

The analysis of OGD Portals points out that there is generally no standardization of datasets among countries in the WB. Each country publishes datasets in different formats. Some of these datasets are considered out of range for assessment based on the 5 star model assessment and examples include those with .zip, .png or .gis. In these cases, we have simply rated them with 1 star level due to the undefined format which means "whatever format in web" out of regular defined formats according to Table 1.

Moreover, the structure of OGD portals among WB countries differentiates. In some portals, datasets are associated with categories while some others use "tags" for grouping and showing the number of available datasets. This makes harder grouping datasets based on categories of organizations. But, they have in common two aspects including a) all portals provide information for organizations that publish data, and b) for each dataset the publishing authority is known.

In addition, the below table shows the number of datasets and organizations that publish open data for each WB country analysed (Table 3).

Table 3. OGD Portals for West Balkan countries

Country	Available datasets	Organizations
Kosovo	205	14
North Macedonia	233	52
Serbia	331	58
Bosna and Herzegovina	305	10
Albania	83	19
Montenegro	106	18

6 Application Prototype

The main components of the application prototype for accessing open data quality, based on the 5 star model, include the Web service and the Quality Assessment (QA) module (Fig. 1). Web service is fetching data from CKAN and DKAN portals using their APIs, while QA module processes data and stores results into the database for further analysis. The Web service is configured to perform calls to OD portals using appropriate APIs on daily basis. This option is configurable and could be adjusted later on if analysis shows that data does not change so frequently on open data portals.

In the application prototype there is not an option to trigger data fetching manually from the Dashboard UI but this is planned as part of the further development. Results are presented to the user in the Dashboard User Interface (UI) designed for this purpose. Analysis component also allows retrieval of historical data on portals and compare different time prints of analysis.

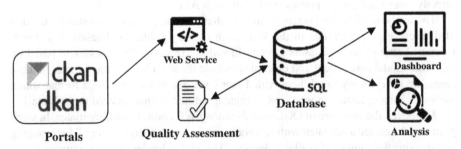

Fig. 1. Application prototype for real time monitoring and assessing OGD

Database ER diagram is shown in Fig. 2. There are four tables in the database used for storing all information necessary for analysis. Table Portals (name/id, url and apiurl), Datasets (id, name, datatype, pushed date and size), Resources (publishing organization, category, dataset url, content type) and Assessment (dataset id, organization id, assessment rate and date). While the first three tables are populated when data is fetched via API, the Assessment table is populated later, after data analysis.

A metric for dataset quality assessment is performed using specific aggregated functions expressed in SQL for grouping and listing datasets based on data formats and then stored in Assessment table. This table is used also for further analysis since it has data stored for different periods of time, which enables data quality analysis for longer periods of time.

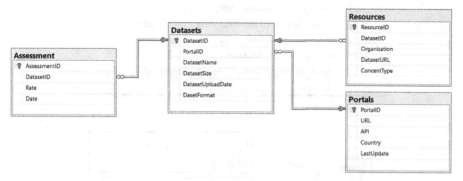

Fig. 2. ER Diagram for storing collected data in SQL Database

A parallel process for fetching data will be used. Apart from storing datasets in a database, additional information during the fetching process will be stored. Table 4 shows undertaken steps during the fetching process.

6.1 Assessing of Dataset Quality

Dataset quality analysis is the main goal of this research. Running web-service will periodically call all APIs of OGD Portals to see if there is any new dataset or new update available. This process is dynamic and cannot be concluded if the quality of datasets is improved right away, until the process finishes. For this purpose, we have used a Dynamic VIEWs in database, which will list all datasets based on their format, including basic information as name of datasets, data format of datasets, organization who published and portal which belong. In Fig. 3 we depicted the role of database VIEWs in dataset assessment.

The calculation of the quality metrics is done as defined in methodology section, while the collection and assessment of general descriptive statistics is done directly after fetching the dataset (see line 12 in algorithm 1.1). The prototype application uses analyzers, so each analyzer has its own task. These analyzers work on the retrieved datasets (and similarly on the resources) and collect and aggregate information. For instance, there is an analyzer for counting the total number for each datatype format

Table 4. Fetching data - steps

Algorithm 1.1 Fetching data from OGD Portals

```
1    def fetch(portals , analyzers ):
2    // downloading dataset store current date
3    snapshot = date.today()
4    for portal in portals :
5    //retrieve list of all datasets within portal
6    datasets_list = get_datasets(portal .api)
7    for id in datasets_list :
8    dataset = get(portal .api , id)
9    // store dataset in database
10   store_in_db(dataset , snapshot)
11   //analyze stored datasets based on formats
12   for analyzer in analyzers :
13   analyzer .analyze(dataset)
14   store_in_db(analyzers . result , snapshot)
15   sleep()
```

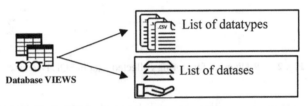

Fig. 3. Dynamic VIEW for listing datatypes and datasets

of datasets and another analyzer for calculating the total number of datasets published by each organization. Regarding data format assessment, according to the 5 star model, grouping of datasets is done based on the following criteria, as given in Table 5.

Table 5. Classification of data formats

Star Rate	★	★★	★★★	★★★★	★★★★★
Datatype Formats	.PDF,.DOC	.XLS,.XLSX	CSV	XML, HTML, JSON	RDF

Also it is important to underline that we found a considerable number of image files on OGD Portals, and these formats are classified as 1 star.

Here we explain how data formats are associated with star rate levels. One star rate means that a dataset is published in whatever format but with an open license. These data formats can be pdf, image or pure text. Two-star rate means if datasets are published as machine-readable data instead of image or table text. These data format can be .XLS or

.XLSX. Three-star rate means that the data is published in the above mentioned formats plus non-proprietary format in .CSV format. Four-star rate is considered if data exists in the above mentioned formats plus it uses open standard form W3C (RDF or SPARQL). These data can be in XML, HTML or JSON format. Five-star rate means if data is available in the above mentioned formats plus link data to other data resources.

Moreover, for calculating dataset quality averages, we have used calculation based on value weight for star rates (1 to 5). For this measurement we have used the following formula which extracts the results based on values:

$$Total\ Score\ Average\ =\ \frac{\sum(1\,star)*1+\sum(2\,star)*2+\sum(3\,star)*3+\sum(4\,star)*4+\sum(5\,star)*5}{Total\ number\ of\ datasets} \quad (1)$$

If we would have a situation where each dataset is published in accordance with the 5 star model, for example, if a dataset exist in .csv format but it also exists in XML or JSON, we rate this datasets with higher star rates. But, in WB countries the situation is different, some datasets exist only in .CSV format but not in other formats or in XLS. Given that, we have applied the formula (1), which evaluates datasets based on the star rate value.

Moreover, in Table 6, we have presented the total score for each country based on the formula (1) calculation, while the total average of all six countries is ~2.5.

Table 6. Average score of data quality assessment

Star rate	★	★★	★★★	★★★★	★★★★★	Total score
Montenegro	0	106	106	106	0	3
Serbia	6	123	109	65	0	2.76
Albania	69	34	66	66	0	2.54
Kosovo	1	204	195	0	0	2.45
North Macedonia	10	178	54	1	0	2.21
Bosna and Herzegovina	147	1	160	0	0	2.04

6.2 Presenting Results in Dashboard

Once fetching process is completed, information on datasets are stored in the database, thus analysis and assessment can start. The application prototype uses integrated dashboard for data presentation and monitoring in real time (Fig. 4). It can be shown in the dashboard how the WB countries stand in terms of dataset quality.

The main page of the dashboard shows some basic information on ODPs, such as the number of available datasets, the number of organizations, information about portals, API used etc. (Fig. 4).

Moreover, dashboard provides possibility to see more details for each country. The details page shows results of dataset quality assessment for selected country (Fig. 5).

Fig. 4. Monitoring portal dashboard

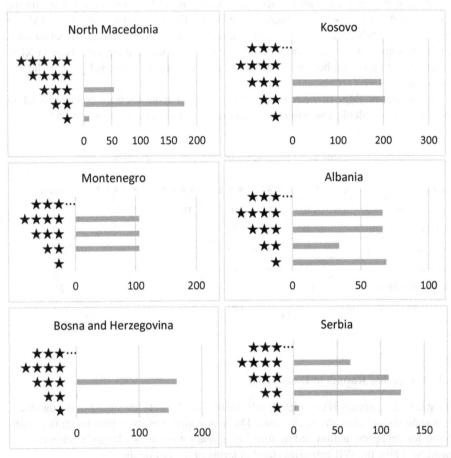

Fig. 5. Dataset quality assessment dashboard

Figure 6 shows comparison of results in terms of datasets quality. This figure provides better overview based on comparative results. It is almost clear for each country what quality of datasets it produces. We can see that BiH followed by Albania are the two countries that have a considerably lower quality of datasets (i.e. datasets are with .pdf or .txt formats). It is easy to notice that none of the analyzed countries have 5 star open data.

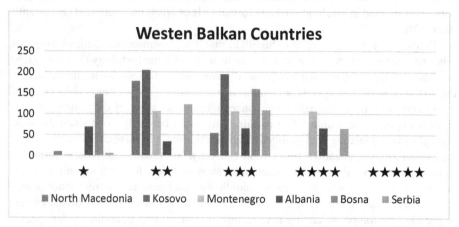

Fig. 6. Comparison of dataset quality assessments

Two stars and three star rates are most common categories for published dataset in all countries. This is shown in Fig. 6 even in Table 6 the results argue this hypothesis since the average of data quality assessment varieties from 2.04 (BiH) and highest average rate 3 (Montenegro). The overall average for 6 WB Countries is 2.5.

Regarding countries characteristics, the results point out that Kosovo uses only two dataset formats for publication (xlsx and csv) so it scores with minimum 2 and maximum 3 stars similar like BiH which has published lot of datasets in PDF, DOC (one star) and CSV (three stars) but no in other data format types. While Montenegro it is the only country which have published all datasets in four different datasets formats (xlsx, csv, xml, json, html) so the results show equivalent bars of graphs for this country due to the equivalent number of datasets for each star rate.

7 Discussion and Future Work

In this research we have presented an application prototype for monitoring open data quality on open data portals in Balkan countries. We have run an analysis to get insights on the current situation of OGD portals including available datasets, resources and dataset quality.

The whole process depends on the running web-service, which is developed to communicate and to retrieve information from APIs offered by OGD portals. Analysis of open data portals in WB countries portrays the existence of two frameworks CKAN and DKAN. So it was necessary to implement APIs calls for both frameworks.

Regarding dataset formats, results shown are specific for each WB country. It is important to highlight that on some ODG Portals (for example Serbia, BiH) a considerable number of impropriate data formats such as .zip, png, and gis exist. These dataset formats cannot be categorized and classified using the 5 star model in higher classes except class one, that simply confirms the availability of such datasets.

The existence of such non-structured formats has an impact on the overall country score results. Moreover, a comparison of assessment results shows that the number of datasets published in machine-readable format (.xls or .xlsx) has increased significantly since 2016.

Furthermore, it is important to highlight that if we compare our methodology with other approaches for assessing data quality using similar methodology (Stojkov 2016), this research approach is considered more dynamic, since the results can change on time due to the prototype designed for the real-time monitoring an assessing of data quality.

Future work on data quality can concentrate on assessing of data quality related to their organization within OGD portals. This could be easily done since the same algorithm for calculating data quality average can be used. First is necessary to find a way for categorizing organizations for example economy, law, finance, and local government, in order to find out data quality assessment average levels for each organization and second to identify which organizations/public agencies produce poorer/higher quality data. In addition, future work could be directed toward advancement of the current application prototype by adding new features for measuring data quality for specific datasets. It could be done using other data quality assessment frameworks based on dimension metrics like data accuracy, data completeness, data consistency, and data timeless.

8 Conclusions

In this paper we have presented a prototype for monitoring and assessing dataset quality in real-time for six WB countries. The analysis of the current situation of OGD portals in WB countries shows that those portals differ in terms of structure, while organizing of datasets is specific for each country.

Results depict the presence of low quality dataset formats as the key concern that affects data reuse. BiH scores the lowest (2.04/5) and Montenegro the highest (3/5), while the other countries do not have a significant difference on values especially Kosovo and Albania (2.45 and 2.54) while Serbia (2.76) is listed as second after Montenegro with (3).

The overall average for all six WB countries is 2.5/5 but it doesn't mean that this value and other country values are permanent. This is due to the dynamic approach and the real-time monitoring done during the first semester of 2020. So, in the future once countries improve their dataset quality formats, the results will also be influenced.

References

1. Blakemore, M., Craglia, M.: Access to public-sector information in Europe: policy, rights, and obligations. Inf. Soc. 22, 13–24 (2006)
2. Chun, S., Shulman, S., Sandoval, R., Hovy, E.: Making connections between citizens, data and government. Inf. Polity 15, 1–9 (2010). In e-Government 2.0
3. Bauer, F., Kaltenböck, M.: Linked open data: the essentials, Vienna (2011)

4. Jetzek, T., Avital, M., Bjorn-Andersen, N.: Data-driven innovation through open government data. J. Theoret. Appl. Electron. Commer. Res. **9**, 100–120 (2014)
5. Chan, C.M.: From open data to open innovation strategies: creating e-services using open government data. In: 46th Hawaii International Conference on System Sciences, Hawai (2013)
6. Berneres-Lee, T.: (2006). http://www.w3.org/DesignIssues/LinkedData.html
7. Whitehouse, O.: Transparency and Open Government (2009). https://obamawhitehouse.arc hives.gov/the-press-office/transparency-and-open-government
8. OGP: Open Government Partnership (2011). https://www.opengovpartnership.org/about/
9. Yannoukakou, A., Araka, I.: Access to government information: right to information and open government data synergy. Procedia Soc. Behav. Sci. **147**, 332–334 (2014)
10. Shadbolt, N., O'Hara, K., Berners-Lee, T., Gibbins, N., Glaser, H., Hall, W.: Linked open government data: lessons from data.gov.uk. Web Internet Sci. IEEE Intell. Syst. **27**(3), 16–24 (2012)
11. Lisowska, B.: Metadata for the open data portals. Technical report. Joined-up Data Standards Project (2016)
12. Bechhofer, S.: Why linked data is not enough for scientists, pp. 599–611 (2013)
13. Dawes, S.S., Helbig, N.: Information strategies for open government: challenges and prospects for deriving public value from government transparency. In: Wimmer, M.A., Chappelet, J.L., Janssen, M., Scholl, H.J. (eds.) Electronic Government. EGOV 2010. Lecture Notes in Computer Science, vol. 6228. Springer, Heidelberg (2010). https://doi.org/10.1007/978-3-642-147 99-9_5
14. Davies, T.: Open data barometer, Global report (2013)
15. Frank, M., Walker, J.: User centred methods for measuring the quality of open data. J. Commun. Inf. **12**(2), 47–68 (2016)
16. Zaveri, A., et al.: Quality assessment methodologies for linked open data. Semant. Web J. **1**, 1–31 (2012)
17. Stojkov, B., et al.: Open government data in western Balkans: assessment and challenges. In: 6th International Conference on Information Society, Technology and Management (ICIST 2016), Belgrade (2016)
18. Batini, A.: Quality assessment and improvement. ACM Comput. Surv. **41**, 16 (2009)
19. Erickson, A., et al.: Open government data: a data analytics approach. IEEE Intell. Syst. **28**, 19–23 (2013)
20. CKAN: CKAN Association, via the Steering Group and Technical Team. https://ckan.org/about/
21. DKAN, "DKAN Open Data Platform https://getdkan.org/
22. A.P. Interface: What is API - Application Program Interface? Webopedia Definition. https://www.webopedia.com/TERM/A/API.html
23. Montenegro, Opendata Montenegro. https://data.gov.me/
24. Bosna and Herzegovina, Opendata.ba. https://opendata.ba
25. North Macedonia Otvorenipodatoci.gov.mk. Добродојдовте. https://otvorenipodatoci.gov.mk/
26. Serbia, Data.gov.rs. Почетак - Отворени подаци. https://data.gov.rs
27. Albania, OpenData - Faqja Kryesore. http://opendata.gov.al/
28. Kosovo, Welcome - RKS Open Data. https://opendata.rks-gov.net/

Analysis of Digitalization in Healthcare: Case Study

Goce Gavrilov[1]([⊠]) [iD], Orce Simov[2], and Vladimir Trajkovik[3] [iD]

[1] University American College, School of Computer Science and Information Technology, Skopje, Macedonia
gavrilovgoce@yahoo.com
[2] Faculty of Computer Science, International Slavic University "Gavrilo Romanovich Derzhavin", Sveti Nikole, Macedonia
osimov@yahoo.com
[3] Faculty of Computer Science and Engineering, University Ss. Cyril and Methodus, Skopje, Macedonia
trvlado@finki.ukim.mk

Abstract. Digitalization in the healthcare system plays a significant role in the improvement of healthcare, as well as in the planning and financing of health services. Health insurance fund's information system is the main component of the information infrastructure that allows assessment of the impact of changes in health care for the population. The main purpose of this paper is to give a brief overview of the impact of the process of digitalization and digital transformation in the Health Insurance Fund at the quality of services. The authors opted for an exploratory study using the digitalization process implemented in the National Health Insurance Fund (Healthcare Fund) and data which were complemented by documentary analysis, including brand documents and descriptions of internal business processes. Digitalization of internal processes and digital transformation of data exchange with other government institutions are main preconditions for establishing business intelligence systems and introducing of e-services. The analysis conducted in this paper shows that the digitalization and digital transformation have a positive impact on the insured people, healthcare providers and companies when fulfilling their administrative obligations and exercising their rights from healthcare. The implemented digitalization processes and digital transformation, presented in this paper, can serve as a valuable input for the healthcare authorities in predictive analytics and making decisions related to healthcare.

Keywords: E-services · Healthcare · Digitalization · Digital transformation · Web portal

1 Introduction

We live in a very exciting time where healthcare does not transform itself but it is transformed by us. The development of medical and information technology is driving better healthcare for humanity across the world although, whereas on the other the costs

© Springer Nature Switzerland AG 2020
V. Dimitrova and I. Dimitrovski (Eds.): ICT Innovations 2020, CCIS 1316, pp. 202–216, 2020.
https://doi.org/10.1007/978-3-030-62098-1_17

are a big concern for all governments, healthcare providers and patients [1]. One of the possible ways to reduce financial expenses and expenditures is to the digitalization of the healthcare sector and paperless communication between the involved parties [2, 3].

Industry development, which includes the emergence and development of the Internet of Things (IoT), sensors, blockchain technology, cloud computing, big data and other innovations, forces organizations to transform traditional business models into a digital model [3]. This transformation involves all aspects of the business starting from the digitization of the processes with the use of digital technologies in products.

Digitalization and digital transformation processes involve changes in the core business operations and modifies products and processes, as well as organizational structures. The companies have to set up management practices to conduct these complex transformations [4].

Thanks to exponential growth in computing power, digital technologies have successfully transformed most sectors of the economy except for healthcare [5]. Despite much money in an investment in digital health technologies by established technology companies [6], healthcare delivery transformation still lags due to numerous reasons. Some of these reasons are the complexity of the healthcare system, high stakes in healthcare, lack of clinical insight for many companies, dissatisfaction and criticism towards Electronic health records (EHRs), the fear of the digital health companies to enter in healthcare regulatory process [6]. The economic gains from the digital transformation still to come and scientists are more convinced of the promise of innovation in healthcare [7].

The application of Industry 4.0 based standard technology, enables the optimization of healthcare processes and resulting in increased efficiency, effectiveness, and overall improvement of clinical quality indicators [4]. It is possible to monitor the healthcare status of the citizen in real-time, after making the digitalization and digital transformation of healthcare processes. Better collaboration and easier communication between patients and healthcare facilities are enabled by digital technologies. However, the main problem in the application and implementation of this kind of technology is primarily related to the problem of securing communication between the devices used, because of threats from cyberspace [7]. Blockchain is one of the promising technologies that offers a solution for secure communication [8, 9]. Reorganizations in healthcare are using very different approaches: from processes' reorganization and skill-mix, staff education and training; appropriate pay and reward systems; to designing new easier digital services. Digitalization and digital transformation provide a new perspective on decision-making processes by creating an infrastructure for the health data and to provide examples for healthcare workers in the healthcare industry using Data mining techniques. Forasmuch as, they facilitate the establishment of the system for data discovery in databases, Data Warehousing (DW), Business Intelligence (BI) [10]. By definition, DW represents a repository of data from many different sources, often in different structures, and is expected to be used under the same combined roof [11, 12]. With digital transformation, the establishment of DW is facilitated [12]. In addition, DW and BI can analyze data from many different sources under the same roof [13].

Many government institutions in Macedonia modernized their information systems and created the basis for online data exchange [12, 14].

In this paper, we present the process and effect of digitalization and digital transformation of some services which are implemented in the Healthcare Fund. The next section presents related work. In Sect. 3, we present a brief description of the architecture of the Healthcare Fund's IT system with a special emphasis on the digitalized processes and digital transformation. Realization of digitalization and digital transformation (case study) is given in Sect. 4, while Sect. 5 presents the evaluation and discussion as well as the novelty of the Macedonian approach. Last section concludes the paper.

2 Related Work

The healthcare costs are declined thanks to the rapid development and advancement of digital technology. Parallel to this, most health economics across the world face to huge challenge in the solving of the problem with the cost rising of healthcare. To solve these challenges, the introduction of digital technology seems more than necessary [15].

2.1 Digitization, Digitalization, Digital Transformation Definition

When we talk for digitalization and digital transformation in healthcare, it is inevitably first to distinguish between the terms digitization, digitalization, digital transformation and the processes that concerned them. The term "digitization" concerns about creating a digital representation of physical objects. Scanning of paper documents and save them as a digital file is an example of the digitization process. In other words, the term digitization means converting something non-digital into a digital representation or artifact. Digitization in healthcare would re-define the patient-doctor relationships and will change the definition of what healthcare is and how it will be delivered and experienced. For future generations, this is the future that guarantees high-quality healthcare.

Digitalization refers to enabling, improving or transforming business processes by leveraging digital technologies (e.g., APIs) and digitized data. This means that digitalization presumes digitization as described previously. Using digital technologies to change a business model and providing new revenue and value-producing opportunities mean the digitalization process. This is the process of moving to digital businesses [16]. Digitalization concerns the use of digital technologies in the context of the production and delivery of a service or product. These digital technologies allow healthcare services to be organized, produced and delivered in new ways. Digitalization is a less technical process (like digitization) than organizational and cultural processes.

The words "digitization" and "digitalization" are a bit different from each other through some people often miss-interpret them. Converting an analog signal to digital like for instance, taking a photograph and then turning it into a digital photograph is the digitalization process [17]. For business, digitalization means an enterprise-wide information system based on the technological foundation of the Internet or related technologies [18].

Many definitions of digital transformation can be found in the literature. Digital transformation concerns the process of full integration of digital technologies into an organization's business [19]. Digital transformation represents a process of exploring digital technologies and supporting capabilities to build a robust, and innovative digital business model to add business values [16]. The concept of digital transformation encapsulates the transformational effect of new digital technologies such as social, mobile, analytical, cloud technologies and the Internet of Things [20]. In some situations, digital transformation is presented as the integration of digital technologies and business processes in a digital economy [21].

Similarly, Gartner Inc. in 2018 [16] introduces a new term for digital transformation as a "Digital business transformation that represents the method of manipulating digital technologies and supporting capabilities to build a robust new digital business model." In this relation, digital transformation shouldn't be confused with Business Process Re-engineering (BRP). BPR focuses on automating noticeably assigned role-based processes while digital transformation obtains and utilizes new data to reimagine these old, rule-based processes and turn them into innovative business models and operations.

2.2 Digitalization of the Healthcare System

Digitization and digital transformation do not only affect the organization for production, but also healthcare organizations, as well as all other service organizations. In the past, the introduction of the healthcare information system (HIS) is considered one of the basic steps of digital transformation. This proved to be one of the smaller parts of the system digitization [3]. More efficient and effective communication in healthcare can result in more economical use of the available capacities [22]. However, through digitalization and digital transformation, the organization does not only adapt to the newly emerging market conditions but also adopts new knowledge that increases its competence. New competence can, in the long run, result in the creation of competitive advantage, but also the increase in the quality of service and products.

The digital transformation in practice is often seen through the possibilities of new technologies in improving internal and external communication [12].

The process of digitalization and digital transformation must begin by identifying the existing situation as well as the context of the organization. Organizations cannot identify the difference between the existing and requested level by users of healthcare services if they aren't defining the context and requirements of stakeholders [7]. The quality of healthcare services depends on the processes of digitalization and digital transformation.

Digitalization and the digital transformation improve public administration functions and business processes in the healthcare area. Also, data governance is one of the essential things for leveraging the data, with the target to provide higher value and achieve greater availability of data for healthcare institutions. So, the digitalization and digital transformation facilitate the realization of the right of health insurance, lead to optimizing costs and provides optimized savings.

3 Architecture of the IT System

One of the basic problems in digitalization and digital transformation flow from the technical point of view in public administration is authentication in the business processes and fulfilling related legal and technical requirements [23, 24]. The acceptance of the concept of existing electronic identity (eID) is crucial for various internet-based services

[25]. Many European countries (like Austria, Belgium, Estonia, Italy, and Finland) had started planning and developing their digitalization processes many years ago. Other countries provide eID via authentication portals using username and passwords (UK and Netherlands) and some countries rely on PKI software certificates and/or base their eID on banking authentication systems (Swedish). The emergence of blockchain technology which found application in the processes of digitalization and digital transformation, is another way to offer authentication mechanisms.

For the analysis covered in this paper centralized IT system with Web portal software solution, implemented in the Healthcare Fund, was used. To support patient's needs in the digitalization process, Healthcare Fund developed an IT system for secure communication between insured people, healthcare providers, companies and Healthcare Fund.

The architecture of the IT system is shown in Fig. 1. The system is based on Microsoft technologies with multi-tier architecture that comprising of:

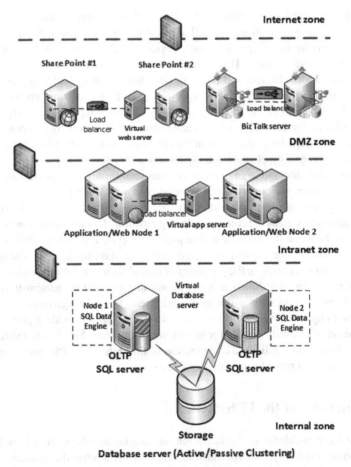

Fig. 1. Architecture of the IT system

– Application/Web servers (exposed to users: employees in the Fund or external users) in Network Load Balancing (NLB) configuration. Application servers also serving as systems for reporting, communicating with SQL DB, integration with external systems via ESB and web services;
– Database servers with installed Microsoft SQL Server cluster;

In the implementation of the system was used: MSMVC (Microsoft Model-view-controller), Web 2.0 standards, single page web applications.

A software solution for the web portal was also implemented as a support of this IT system. It consists of the following servers:

– SharePoint Portal servers in NLB configuration,
– BizTalk servers in NLB configuration.

Web portal software solution with Enterprise Service Bus (ESB) based on BizTalk is upgrading the existing IT system. BizTalk Server serves as a broker for integration with the external environments. This Web portal enables insured people to communicate and cooperate with their doctors and Healthcare Fund easily and securely. Web portal software solution uses publish/subscribe model of communication with BizTalk Server. Also, the BizTalk Server is integrated across the service level of existing systems on the other side. Communication will always be in the direction from the portal towards ESB, ESB towards existing systems and vice versa. Service-oriented architecture (SOA) is enabled from the ESB which is the basic architecture of the solution. The implementation of BizTalk as ESB enables modular and component-based architecture. The authentication of the users is realized through the integration services with Clearing house and Macedonian Telekom as registered issuers of qualified certificates in Macedonia.

4 Digitalization of Health Insurance Fund: Case Study

Electronic service - proof of paid contribution is used in this paper as an example of digitalization and digital transformation. This electronic service is an example of a service for administrative relief for the companies, healthcare facilities, and insured people. A brief description of it is given in the text below.

4.1 Electronic Service - Proof of Paid Contribution

Healthcare Fund, according to the legal regulations, leads and manages with a register of bonds for paid contributions and registers for the calculation and contributions' payment. Bonds for the calculation and payment of contributions for health insurance do registration in health insurance for insured people in the Fund's IT system. Also, other governmental institutions that are obliged to pay contributions for insured people, do registration in health insurance for their categories. As proof of paid contributions for healthcare insurance, in the past, the Healthcare Fund was issuing a paper - based certificate for paid contribution so-called "blue coupons".

Digitalization of internal business processes that cover the process of registration in healthcare insurance, payment, and control of contribution, implementation of digital transformation in the processes of data exchange with external institutions, automatic processing of data, check od identity, offer new possibilities to the representation of these services. So, new services like electronic data exchange between concerned institutions, implementation of electronic registration and deregistration in health insurance, were implemented. This, from the other side, imposed the need from the introduction of the electronic service - checking of insurance status. To allow real-time checking of insurance status, there are background processes of automatic matching of data from the Fund's database with data from other institutions. Healthcare Fund, for the needs of this service, exchanges data with (see Fig. 2):

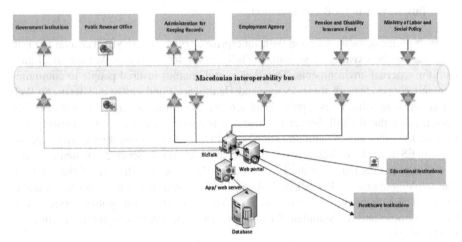

Fig. 2. Electronic matching data from different sources

- Employment Agency (EA). EA has the data on the employment of persons who are also insured, people. EA sends the information for each M1-registration, M2-deregistration, and Mx-change. The exchange of data between Fund and EA is performed by using the so-called "push" methodology every 15 min.
- Public Revenue Office (PRO). PRO has the information about all active bonds, the changes of active bonds, as well as data for all newly formed bonds, data for individual payers for health insurance, data for each insured person in compulsory social insurance. The exchange of data between the Healthcare Fund and PRO is done twice per day via the SFTP server with the encrypted files.
- Pension and Disability Insurance Fund (PDIF). PDIF has the information for new pensioners and data for pensioners with "pause status". The exchange of data between Healthcare Fund and PDIF is realized by "request" from web service on the Fund's side to the web service on the side of PDIF.
- Administration for Keeping Records (AKR) - AKR has the information for newborns, marriage, and death certificates. The exchange of data between Fund and AKR is

realized by "request" from web service on the Fund's side to the web service on the side of AKR for newborns, marriage certificates and push methods for death certificates in reverse.

- Educational Institutions (Universities and Faculties) - (EIs). EIs have the information for all full-time students who have enrolled in the current semester. Receiving data from EIs is via a web portal from the authorized persons from the faculties, two times per year.
- Ministry of Labor and Social Policy (MLSP) - Ministry of Labor and Social Policy (MLSP) - MLSP has the data for social categories of citizens. MLSP sends (real-time) the data for each registration, deregistration, and change of these categories and information for a certificate of participation exemption in healthcare services.
- Companies - An authorized person from the companies, send (via a web portal) the information for registration of new members in healthcare insurance or member deregistration.

The data exchange with other government institutions is happening across the inter-operability framework of the Ministry of information society and administration called Macedonian interoperability bus (MIS).

Some of these data are imported in real-time others at specific time intervals. Several times per day, the imported data into the Fund's database, are processed, compared and matched, the data for paid contributions and data obtained from other institutions with the data in the Fund's database. Also, several times per day the recalculation procedure for the expiration date of insurance for each insured person is running. All of these operations are performed to provide accurate data for the validity of health insurance or servicing of e-service - proof of paid contribution.

The background processes are the basis for the implementation of e-service - proof of paid contribution. This service is implemented on the Fund's web portal, available for health facilities via Web Service through the Fund's ESB, Web Service through MIS for other government institutions. Access to information from this service is available to external users only by digitally signed request with the digital certificate or via web service across MIS of Ministry of information society and administration.

5   Evaluation and Discussion

In previous section we gave a brief description of one e-service offered by the Healthcare Fund (Electronic service - proof of paid contribution). Although we have described one service, however, the synchronization processes and data matching that are occurred to serve this service, offer additional services for healthcare facilities, companies, organizations and insured people. In this section, we will make a comparison between using this service (and additional services) before and after the digitalization process and digital transformation. Within the evaluation of the proposed new e-services, we will determine the advantages of digitalization and digital transformation versus conventional paper-based services. Also, we will determine what is the better service, conventional paper-based service or digitalized e-service?

In the evaluation of implemented digitalization and digital transformation, the following parameters are taken into account:

- valorization of time spent filling the requested documents in Healthcare Fund for exercising the rights of healthcare insurance and services;
- savings in the form of petrol and parking fees for acquiring the requested documents, issuing some documents, etc.
- savings in the form of material costs

In the past, as proof of paid contributions for health insurance, Healthcare Fund issued the proves for paid contributions so called "blue coupons". "Blue coupons" have been printed for all insured people (carriers of health insurance), except for pensioners, every month. According to Healthcare Fund [26], the number of these insured people (carriers at healthcare insurance) in the last few years on average is about 854.661 per year, so the total number of printed "blue coupons" for 12 months in average is 10.255.936. Our calculations (for "blue coupons") for procurement of paper, ribbons and amortization and procurement of equipment, are amounting to 164.358 € per year (as shown on Fig. 3) or for last six years (2014–2019) after finishing the digitalization is about 986.000 € (Table 1).

Table 1. Savings from the printing

Years	2016	2017	2018
Carrier in HC for printing BC	850.433	860.347	853.204
Total BC per year	10.205.196	10.324.164	10.238.448
Total prices per year (in €)	163.545	165.451	164.078

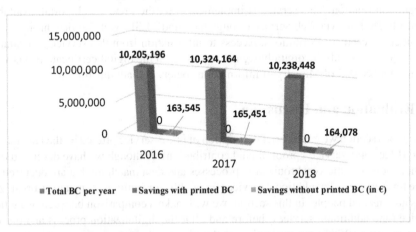

Fig. 3. Savings from the printing blue coupons for three years

According to the Employment Agency [27], the number of newly registered employments (M1 registration) for the last three years is around 395.786 (see Table 2). This means that every year in the Healthcare Fund is realized approximately such a number of registrations in health insurance for employed people and members insured through them. So, every year there are about 395.786 of visits in branch offices from companies' representatives only on the basis of registration in health insurance.

Table 2. Number of registration in healthcare of new employments

Year	2017	2018	2019
No of M1 registration	370.474	410.716	406.168

If we suppose that for every coming to Fund's branch offices are only taken into account only the cost for petrol and parking amounts (average of 2 €), then the total savings for companies per year is around 792.000 € (as shown on Fig. 4) or for last six years (2014–2019) after finishing the digitalization is about 4.749.432 €.

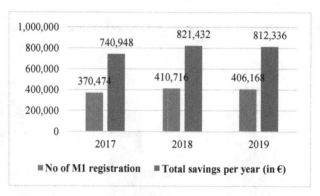

Fig. 4. Savings for the companies

According to the State Statistical Office [28], the number of newborns children in recent years is around 22.030 children per year (21.333, 21.754, 23.002 for 2018, 2017, 2016 respectively). If for the registration at health insurance is not required paper birth certificates (as in our case after digitalization), that according to [29] one birth certificate costs 2,44 €, then regarding on this basis citizens would save around 53.000 € per year (as shown on Fig. 5) or for last six years (2014–2019) after finishing the digitalization is about 322.514€.

According to the State Statistical Office [30], the number of full-time students in recent years is around 50.348 full-time students every year. If for extension of health insurance for full-time students is not required paper-based certificates (as in our case after digitalization), we have savings. So that, according to the price for this certificates from some faculties, one paper-based certificate costs 2,44 €, then regarding on this basis citizens would save around 122.000 € per year (as shown on Fig. 6) or for last six years (2014–2019) after finishing the digitalization is about 737.090 €.

Previously mentioned savings are enabled after the implemented process of digitalization and digital transformation.

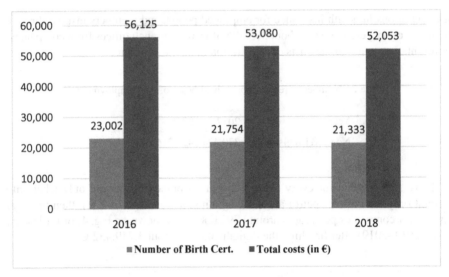

Fig. 5. Savings by digitalization of the paper based birth certificates

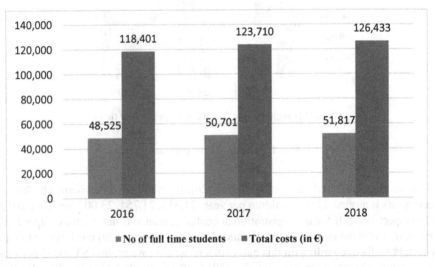

Fig. 6. Savings from digitalization of the paper-based certificate for full-time student

With these processes of digitization and digital transformation, insured people, healthcare providers and companies will make savings in time and transportation. The institutions will make savings in the form of paper, printing expenses and labor. Savings are made as a result of the elimination of requested paper documents, which will be exchanged between the Healthcare Fund and the other institutions in the new digital procedure. Despite the obvious savings for Fund because of reduced paperwork, an assumption is being introduced that the new digital work process will release significant administrative labor, which in the future can be redistributed to other work.

An additional aspect in calculating savings in paper, for all involved institutions, is to convey these savings into environment protection, as a result of reduced demand for paper. Finally, the time spent on the achievement of rights of health insurance, the healthcare services is invaluable and for it we have not made a calculation.

6 Conclusion

The purpose of this paper is to study and analyze the effect of digitalization and digital transformation (such as information quality impact, services quality impact or organizational impact of the whole healthcare system). In this paper, we have described the results from the digitalization process accompanying with digital transformation on the service proof of paid contribution. We have evaluated the benefit of these processes, applied to one service (proof of paid contribution) by calculating the savings for the 6 years after the implementation of this service as well as some associated services to this.

The analysis conducted in this paper shows that the digitalization of processes in Healthcare Fund would have a positive impact on the insured people, healthcare providers, and companies when fulfilling their administrative obligations and exercising their rights. But digitalization of Fund's services will impose side effects that should not be neglected. These side effects are the reduction of revenues for issuing documents from other institutions and savings for insured people and institutions because of simplified procedures.

Insured people, healthcare facilities, and companies will benefit because services (main service and the additional services, described in this paper) are provided in one place (from their offices), which brings savings in terms of time and transportation. We can summarize a few potential benefits and losses of digitalization and digital transformation implemented on one service in the Healthcare Fund. Healthcare Fund and other institutions will have fewer costs for printing because of reduced issuing of documents to insured people. The increased efficiency of the employees will mean that more time can be devoted to other work tasks. Providing simplified and more efficient digitalized services will enable faster data access and more accurate data in terms of timely updates of data from the electronic data exchange. Also, providing simplified and more efficient digitalized services will show an improved image of the Healthcare Fund for their determination towards better quality service. Insured people will save time by going to fewer places to obtain healthcare services, which will reduce the costs of transportation. Last but not least, the reduction of the printed paper defines the positive impact of a greener society.

According to the results from the analysis (available data and presented graphics), all users of the service-paid contributions and additional services will have financial benefits and savings in unnecessary spending time. The analysis presented in this paper can serve as a valuable input for the healthcare authorities in making decisions related to digitalization and digital transformation in healthcare. As a leader in healthcare in Macedonia, Healthcare Fund began and will continue the digitalization process and digital transformation. But attaining and maintaining the trust of insured people and the other stakeholders is crucial for success.

Digital transformation of healthcare processes comes not only to the implementation of new technologies, but organizations also have the opportunity to improve existing processes which will influence the increase in the quality of health care provided. Higher quality of health services results in a higher quality of life for its users. By implementing digitalization and digital transformation of the healthcare system, the health organization affects the satisfaction of the user with the service and increase the quality of life of the user. But, the evaluation of digital transformations is a long-standing activity in both research and practice. Given its extensive history, the persistent problems and lack of progress in the field are surprising–even alarming.

References

1. Global Burden of Disease Health Financing Collaborator Network: Future and potential spending on health 2015–40: development assistance for health, and government, prepaid private, and out-of-pocket health spending in 184 countries. Lancet **389**, 2005–2030 (2017). https://doi.org/10.1016/S0140-6736(17)30873-5
2. Bria, B., Kennedyc, R., Womack, D.: Learning from the EHR for quality improvement, a descriptive study of organizational practices. J. Healthc. Inf. Manage. **26**, 46–51 (2012)
3. Zunac A.G., Kovacic, M., Zlatic, S.: The impact of digital transformation on increasing the quality of healthcare. In: 7th International Conference on Quality System Condition for Successful Business and Competitiveness, pp. 49–56. Association for Quality and Standardization of Serbia (2019)
4. Matt, C., Hess, T., Benlian, A.: Digital Transformation Strategies. Bus. Inf. Syst. Eng. **57**(5), 339–343 (2015). https://doi.org/10.1007/s12599-015-0401-5
5. Walsh, N.M., Rumsfeld, J.S.: Leading the digital transformation of healthcare: the ACC innovation strategy. J. Am. Coll. Cardiol. **70**(21), 2719–2722 (2017). https://doi.org/10.1016/j.jacc.2017.10.020
6. The wonder drug: a digital revolution in healthcare is speeding up. The Economist (2017). https://www.economist.com/news/business/21717990-telemedicine-predictive-diagnosticsw earable-sensors-and-host-new-apps-will-transformhow. Accessed 21 Apr 2020
7. Burton-Jonesa, A., Akhlaghpoura, A., Ayreb, S., Bardec, P., Staibd, A., Sullivane, C.: Changing the conversation on evaluating digital transformation in healthcare: insights from an institutional analysis. Inf. Organ. **30**(1) (2020). https://doi.org/10.1016/j.infoandorg.2019.100255
8. Peterson, K.J., Deeduvanu, R., Kanjamala, P., Boles, K.: A blockchain-based approach to health information exchange networks. In: Proceedings of the NIST Workshop Blockchain Healthcare, vol. 1, pp. 1–10 (2016)
9. Alla, S., Soltanisehat, L., Tatar, U., Keskin, O.: Blockchain technology in electronic healthcare system. In Barker, K., Berry, D., Rainwater, C. (eds.) In: Proceedings of the 2018 IISE Annual Conference. Institute of Industrial & Systems Engineers (IISE), vol. 1, pp. 754–760 (2018)
10. Cifci, M.A., Hussain, S.: data mining usage and applications in health services. Int. J. Inf. Visual. **2**(4), 225–231 (2018). https://doi.org/10.30630/joiv.2.4.148
11. Raghupathi, W.: Data mining in healthcare. Healthcare Informatics: Improving Efficiency through Technology, Analytics, and Management, pp. 353–372 (2016). https://doi.org/10.1201/b21424
12. Gavrilov, G., Vlahu-Gjorgievska, E., Trajkovik, V.: Healthcare data warehouse system supporting cross-border interoperability. Health Inf. J. **1**(12) (2019). https://doi.org/10.1177/1460458219876793

13. George, J.L., Kumar, B.P.V., Kumar., V.S.: Data warehouse design considerations for a healthcare business intelligence system. In: Proceedings of the World Congress on Engineering WCE 2015, London, U.K (2015)

14. Todevski, M., Sarkanjac, S.J., Trajanov, D.: Analysis of introducing one stop shop administrative services: a case study of the Republic of Macedonia. Transylvanian Rev. Admin. Sci. **38**, 180–201 (2013)

15. Ravish, G.: Digitization of healthcare solutions-an enabler of health and disease management. Int. J. Res. Eng. IT Soc. Sci. **8**(12), 178–180 (2018)

16. Gartner Inc. 920180: Gartner IT Glossary. https://www.gartner.com/en/information-techno logy/glossary. Accessed 10 April 2020

17. Schallmo, D.R.A., Williams, C.A.: Digital transformation of business models. In: Schallmo, D.R.A., Williams, C.A. (eds.) Digital Transformation Now! Guiding the Successful Digitalization of Your Business Model, pp. 9–13. Springer, Heidelberg (2018). https://doi.org/10. 1109/MS.2018.2801537

18. Li, J., Merenda, M., Venkatachalam, A.R.: Business process digitalization and new product development: an empirical study of small and medium-sized manufacturers. E-Bus. Appl. Prod. Dev. Competitive Growth **5**(1), 49–64 (2009). https://doi.org/10.4018/978-1-60960-132-4.ch003

19. Khanboubi, F., Boulmakoul, A.: State of the art on digital transformation: focus on the banking sector. In: International Conference on Innovation and New Trends in Information Systems, pp. 9–19 (2018)

20. Ebert, E., Duarte, C.H.C.: Digital Transformation. IEEE Softw. **35**(4), 16–21 (2018). https://doi.org/10.1109/MS.2018.2801537

21. Liu, D.Y., Chen, S.W., Chou, T.C.: Resource fit in digital transformation: lessons learned from the CBC Bank global e-banking project. Manag. Decis. **49**(10), 1728–1742 (2011). https://doi.org/10.1108/00251741111183852

22. AlQarni, Z.A., Yunus, F., Househ, M.S.: Health information sharing on Facebook: An exploratory study on diabetes mellitus. Journal of Infection and Public Health **9**(6), 708–712 (2016). https://doi.org/10.1016/j.jiph.2016.08.015

23. Kutylowski, M., Kubiak, P.: Polish concepts for securing e-government document flow. In: Pohlmann, N., Reimer, H., Schneider, W. (eds.) Proceedings of the ISSE 2010 Securing Electronic Business Processes, pp. 399–407. Springer Fachmedien Wiesbaden GmbH, Berlin (2011). https://doi.org/10.1007/978-3-8348-9788-6_39

24. Gavrilov, G., Simov, O., Trajkovik. V.: Blockchain-based model for authentication, authorization, and immutability of healthcare data in the referrals process. In: Proceedings of the 17th International Conference on Informatics and Information Technologies (2020)

25. Leitold, H., Zwattendorfer, B.: STORK: architecture, implementation and pilots. In: Pohlmann, N., Reimer, H., Schneider, W. (eds.) Proceedings of the ISSE 2010 Securing Electronic Business Processes, pp. 131–142. Springer Fachmedien Wiesbaden GmbH, Berlin (2011). https://doi.org/10.1007/978-3-8348-9788-6_13

26. Health Insurance Fund of Macedonia: Annual report for 2018. http://www.fzo.org.mk/WBS torage/Files/Godisen%202018%20KONECNO.pdf. Accessed 10 April 2020

27. Employment Agency: Annual report for 2017. https://av.gov.mk/godishni-izveshtai.nspx. Accessed 29 Apr 2020

28. State Statistical Office of Republic of North Macedonia: Birth rate in the Republic of Macedonia in 2016, 2017, 2018. State Statistical Office of Republic of North Macedonia, Skopje (2016, 2017, 2018). http://www.stat.gov.mk/PrethodniSoopstenijaOblast.aspx?id=8& rbrObl=2. Accessed 5 Apr 2020

29. Procedure for issuing a birth certificate. https://akademik.mk/servisni-informacii/postapka-za-izdavane-izvod-od-matichna-kniga-na-rodeni/. Accessed 5 Apr 2020
30. State Statistical Office of Republic of North Macedonia: Enrolled students in undergraduate studies in the academic year 2016/2017, 2017/2018, 2018/2019. State Statistical Office of Republic of North Macedonia, Skopje (2017, 2018, 2019). http://www.stat.gov.mk/Publikaci iPoOblast.aspx?id=38&rbrObl=5. Accessed 10 Apr 2020

Correlating Glucose Regulation with Lipid Profile

Ilija Vishinov[1]([✉]) [iD], Marjan Gusev[2] [iD], Lidija Poposka[3] [iD], and Marija Vavlukis[3] [iD]

[1] Innovation Dooel, Skopje, North Macedonia
ilija.vishinov@innovation.com.mk
[2] Faculty of Computer Science and Engineering,
Ss. Cyril and Methodius University, Skopje, North Macedonia
marjan.gushev@finki.ukim.mk
[3] Clinic of Cardiology, Ss. Cyril and Methodius University,
Skopje, North Macedonia
lidijapoposka@gmail.com, marija.vavlukis@gmail.com

Abstract. *Objectives*: The goal of this research was to detect the glucose regulation class by evaluating the correlation between the lipid profile of patients and their glucose regulation class.

Methodology: The methods used in this research are: i) Point Biserial Correlation, ii) Univariate Logistic Regression iii) Multivariate Logistic Regression iv) Pearson Correlation and v) Spearman Rank correlation.

Data: The dataset consists of the following features: age, BMI, gender, weight, height, total cholesterol (Chol), HDL cholesterol (HDL-C), LDL cholesterol (LDL-C), triglycerides (TG), glycated hemoglobin (HbA1C), glucose regulation and diabetes classes, history of diabetes, heart and other chronic illnesses, habitual behaviors (smoking, alcohol consumption, physical activity), and medications intake (calcium channel blockers, BETA blockers, anti-arrhythmic, AKE/ARB inhibitors, diuretics, statins anti-aggregation medication and anticoagulants).

Conclusion: The methodologies that were worked through with our data in search for correlations of the lipid profile with HbA1c or the glucose regulation classes gave some significant correlations. Regarding the glucose regulation classes W and B the methods showed statistically significant negative correlations with Chol, HDL-C and LDL-C. When it comes to the correlations of the lipid profile with HbA1c, for all patients there were significant negative correlations with Chol (corr $= -0.264$, p $= 0.002$), LDL-C (corr $= -0.297$, p < 0.001) and HDL-C (corr $= -0.28$, p $= 0.001$) and a significant positive correlation with TG (corr $= 0.178$, p $= 0.03$). The correlations mentioned are the stronger ones that were found for linear relationships. For non-diabetic patients there was a stronger positive non-linear correlation for HbA1c and HDL-C (corr $= 0.511$, p $= 0.006$), and a slightly weaker linear correlation (corr $= 0.393$, p $= 0.043$). For prediabetic patients there were no significant correlations. For type 2 diabetes stronger significant negative non-linear correlations were found for HbA1c with LDL-C (corr $= -0.299$, p $= 0.023$) and HDL-C (corr $= -0.438$, p $= 0.001$). The linear relationships

© Springer Nature Switzerland AG 2020
V. Dimitrova and I. Dimitrovski (Eds.): ICT Innovations 2020, CCIS 1316, pp. 217–227, 2020.
https://doi.org/10.1007/978-3-030-62098-1_18

were again, slightly weaker with LDL-C (corr = -0.273, p = 0.038) and with HDL-C (corr = -0.391, p = 0.002).

Keywords: Cholesterol · Lipid profile · Glucose regulation · Diabetes

1   Introduction

This is a followup research [24] within the Glyco project [13], which aims at detecting correlation between electrocardiograms and blood glucose levels.

The goal of this research is to evaluate the correlation between the lipid profile and the glucose regulation classes or the blood glucose levels of patients. This motivation which led us to conduct it, is the possibility of complementing the diabetic record of the patient with information inferred from the patients' lipid profile, aside from that of the electrocardiograms. This opens the possibility of making more accurate predictions about the their diabetic state.

The private database of clinical research within the project Glyco [13] consists of 161 patients with records containing values about the patients' age, height, weight, BMI, lipid profile (Chol, HDL-C, LDL-C, TG), glycated hemoglobin levels with respective glucose regulation and diabetes classes, history of diabetes, heart and other chronic illnesses, habitual behaviors (smoking, alcohol consumption, physical activity) and medications intake (calcium channel blockers, BETA blockers, antiarrhythmics, AKE/ARB inhibitors, diuretics, statins, antiaggregation medication and anticoagulants).

Four continuous measures: Total Cholesterol (Chol), HDL cholesterol (HDL-C), LDL Cholesterol (LDL-C) and Triglycerides (TG) determine the lipid profile. The measured Glycated Hemoglobin (HbA1c) is used to determine the blood glucose.

The paper adheres to the following structure: Sect. 2 gives an overview of the related work and Sect. 3 describes the experiments. Results of correlation are presented in Sect. 4 and discussed in Sect. 5.

2   Related Work

Several research papers reveal their results about correlation of the lipid profile and the glucose regulation class. Vinodmahato et al. [23] found a significant positive correlation between HbA1c and Chol, LDL-C and TG but a non-significant positive correlation between HbA1c and HDL-C.

Yan et al. [25] only found a significant correlation, a positive one with LDL-C. The other correlations were non-significant. Bener et al. [1] found highly significant higher levels of Chol, LDL-C and TG and lower HDL-C levels in patients with diabetes.

In our previous research [24] we give an overview of several related research papers: for pre-diabetes patients [2], type 1 diabetes patients [16,18], type 2 diabetes patients [2,3,6–9,11,12,14,15,17,20], and [10]. These papers conclude a positive correlation between the non-HDL-C (LDL-C, Chol and TG) and HbA1c except for one exception, Kim et al. [16] who found a non-significant correlation with the LDL-C.

The correlations with HDL-C were split almost evenly between having a statistically significant negative correlation and having a non-significant correlation.

Table 1. Glucose regulation classes based on diabetes type of the proprietary dataset (Innovation Dooel)

ID	Regulation class	Diabetes class	HbA1c (%)	Distribution (%)
N	Normal regulation	Non-diabetic	≤5.6	21.1
I	Impaired regulation	Prediabetic	≤6.4	31.7
O	Optimum regulation	Type 2 diabetic	≤7.4	19.3
M	Medium regulation	Type 2 diabetic	≤9.4	19.9
P	Poor regulation	Type 2 diabetic	>9.4	8.1

Table 2. Binary glucose regulation classes of the proprietary dataset (Innovation Dooel)

ID	Regulation class	HbA1c (%)	Distribution (%)
W	Well regulation	≤6.4	52.8
B	Bad regulation	>6.4	47.2

3   Experimental Setup

The dataset consists of 161 records of unique patients. Their age distribution was 60.23 ± 10.49 years (mean $\pm$ standard deviation).

The lipid profile consists of these 4 features followed by their mean $\pm$ standard deviation:

- Chol - Total Cholesterol, 5.21 ± 1.23 mmol/L.
- HDL-C - High-Density Lipoprotein Cholesterol, 1.20 ± 0.35 mmol/L.
- LDL-C - Low-Density Lipoprotein Cholesterol, 3.06 ± 1.11 mmol/L.
- TG - Triglycerides, 1.96 ± 1.23 mmol/L.

Besides the continuous to continuous correlations, we will assess the correlation for categorical to continuous features with the appropriate methods. The classes which belong to the glucose regulation type, is what we will correlate to. The patients are classified based on two criteria dependent on HbA1c. The first type as shown in Table 1 classifies the patients without diabetes as with normal regulation (N), patients with prediabetes as with impaired regulation (I) and patients with diabetes in three ordinal classes, with optimal regulation (O), medium regulation (M) and poor regulation (P). The second classification showed in Table 2, regardless of whether or not patients had diabetes, considers if the glucose is well regulated (W) or the glucose regulation is bad (B).

The research is based on the following correlation methods compliant to the research goal, assuming that one of the analyzed parameters is a continuous feature, and the other a discontinuous (binary or with multiple classes):

- *Point-Biserial correlation (PBC)*. Assumptions that need to be met when applying this method are homo-scedasticity between the features i.e. the variance of the continuous feature is not significantly different when divided based on the binary feature. Also the continuous feature should have a gaussian or a very closely gaussian-like distribution. The output (result) is a point biserial correlation coefficient and a p value given the hypothesis that there is no correlation.
- *Univariate Logistic Regression (LR)*. A method that is less demanding regarding the assumptions. Given that there is only one predictor (independent) feature, the assumption space reduces even further eliminating the need for reducing multicollinearity and linear relationship to the log odds. If a correlation exists, we should be able to construct an accurate, precise and sensitive classifier.
- *Multinomial Logistic Regression (MLR)*. With multinomial logistic regression we tried to construct a classifier directly for the multiclass problem and seeing if any combination of the cholesterol features can give a highly accurate classifier.

When correlating HbA1c to the lipid profile the following methods were used:

- *Pearson Correlation*. A correlation method that captures linear correlations between two continuous features.
- *Spearman Rank Correlation*. A correlation method that captures both linear and non-linear monotonic relationships between two continuous features.

The following tools were used in evaluation and making decisions:

- *PBC coefficient and p-value* were evaluated based on the point-biserial correlation coefficient which ranges from -1 to 1 (1 and -1 indicating perfect correlation but with different directions and 0 indicating no correlation) and the associated p-value for the null hypothesis that the correlation is 0 [4]. Considering the acceptance threshold of the p-value, anything in the close upper proximity to 0.05 should be taken very cautiously and anything below that we will consider to be statistically significant and reject the null hypothesis. [22].
- *Classificator evaluation for the logistic regression*. The ROC (receiver operating characteristic) and PR (precision-recall) curves and their corresponding

AUC's (area under curve) were used. The abbreviations TP, FP, FN and TN represent true positives, false positives, false negatives and true negatives respectively. The ROC curves plot the true positive rate (TP/(TP+FN)) against the false positive rate (FP/(FP+TN)) and the PR curve plots the precision (TP/(TP+FP)) against the recall (TP/(TP+FN)). The AUC value of the plots summarizes the quality of the logistic regression models.

- *PR AUC and cross-validation for the logistic regression.* Because of the imbalanced dataset when dealing with the dichotomous features for the regulation classes, accuracy as a measure can be misleading when evaluating the model. Thus, ROC curves, PR curves and their corresponding areas under the curve mentioned in the previous paragraph were used. More weight was given to the PR AUC, because PR curves better capture the quality of the classifier when dealing with an imbalanced set [5, 19, 21]. The ROC and PR curves were evaluated over 10-fold stratified cross-validation, providing a mean and variance of the scores. Since scores equal to 0.5 corresponds to a classifier giving random guesses and a score of 1 corresponds to a perfectly precise and sensitive classifier, we let them represent each end of the correlation spectrum, i.e. no correlation and perfect correlation respectively.

- *Multinomial Logistic Regression.* This model was evaluated based on accuracy only.

Table 3. Point-Biserial correlations between cholesterol and glucose regulation classes for which homoscedasticity is satisfied. Significant correlation (p ≤ 0.05) values are labeled with an asterisk (*). The missing combinations are due to the unsatisfied homoscedasticity condition.

Cholesterol	reg.class	PBC	p-value	reg.class	PBC	p-value
Chol	I	0.129	0.142	W	−0.217*	0.013
Chol	N	0.122	0.164	B	−0.217*	0.013
Chol	M	−0.158	0.072			
Chol	P	−0.18*	0.039			
LDL-C	N	0.117	0.19	W	−0.249*	0.005
LDL-C	I	0.161	0.072	B	−0.249*	0.005
LDL-C	O	0.005	0.953			
LDL-C	P	−0.213*	0.016			
HDL-C	N	0.076	0.388	W	−0.212*	0.015
HDL-C	I	0.118	0.182	B	−0.212*	0.015
HDL-C	O	0.108	0.219			
HDL-C	M	−0.173*	0.049			
TG	I	−0.114	0.192			
TG	O	0.015	0.859			
TG	M	0.107	0.222			

4 Results

The PBC correlations and their associated p-values for the combinations of cholesterol and glucose regulation class for which homoscedasticity is satisfied are presented in Table 3.

The results of conducting the LR method are presented in Table 4. LR was used to construct a classifier with the cholesterol attributes as individual predictors for the diabetes feature. In the following tables, the ROC and precision-recall AUCs are presented for each cholesterol feature.

Table 4. Glucose regulation classes inferred by cholesterol using logistic regression

Cholesterol	reg.class	ROC AUC	PR AUC	Cholesterol	reg.class	ROC AUC	PR AUC
Chol	N	0.56 ± 0.19	0.40 ± 0.19	HDL-C	N	0.43 ± 0.29	0.38 ± 0.28
Chol	I	0.57 ± 0.18	0.53 ± 0.18	HDL-C	I	0.58 ± 0.19	0.55 ± 0.20
Chol	O	0.36 ± 0.11	0.20 ± 0.02	HDL-C	O	0.55 ± 0.22	0.29 ± 0.16
Chol	M	0.58 ± 0.23	0.44 ± 0.31	HDL-C	M	0.64 ± 0.19	0.48 ± 0.24
Chol	P	0.67 ± 0.30	0.42 ± 0.36	HDL-C	P	0.71 ± 0.25	0.39 ± 0.32
Chol	W	0.61 ± 0.15	0.78 ± 0.11	HDL-C	W	0.61 ± 0.23	0.78 ± 0.15
Chol	B	0.61 ± 0.15	0.63 ± 0.19	HDL-C	B	0.61 ± 0.23	0.64 ± 0.20
LDL-C	N	0.62 ± 0.22	0.46 ± 0.24	TG	N	0.55 ± 0.20	0.39 ± 0.26
LDL-C	I	0.58 ± 0.14	0.58 ± 0.13	TG	I	0.54 ± 0.19	0.50 ± 0.21
LDL-C	O	0.34 ± 0.17	0.19 ± 0.02	TG	O	0.40 ± 0.21	0.22 ± 0.04
LDL-C	M	0.60 ± 0.17	0.41 ± 0.13	TG	M	0.59 ± 0.25	0.45 ± 0.23
LDL-C	P	0.73 ± 0.29	0.41 ± 0.31	TG	P	0.63 ± 0.27	0.34 ± 0.36
LDL-C	W	0.63 ± 0.11	0.77 ± 0.10	TG	W	0.60 ± 0.18	0.76 ± 0.16
LDL-C	B	0.63 ± 0.11	0.64 ± 0.15	TG	B	0.59 ± 0.18	0.64 ± 0.16

Reevaluation of the correlation between the cholesterol attributes and HbA1c with calculation of the Pearson and Spearman coefficients are presented in Table 5 which are a further evaluation on the results from our previous research [24]. Additionally in this paper we evaluate the correlation coefficients for each diabetes type that is present in our data.

Results for MLR method are shown in Table 6. The multi-class classifier was attempted for each feature in the lipid profile, then all combinations of two, three and four.

Table 5. Pearson and Spearman coefficients for HbA1c with each cholesterol feature for each diabetes type. Highly significant correlations (p ≤ 0.005) are labeled with '**'. Significant correlations (p ≤ 0.05) are labeled with '*'.

Diabetes type	Cholesterol	Glucose	Pearson c.	p-value	Spearman c.	p-value
All patients	Chol	HbA1c	−0.264**	0.002	−0.225*	0.009
	LDL-C	HbA1c	−0.297**	0.001	−0.271**	0.002
	HDL-C	HbA1c	−0.280**	0.001	−0.225*	0.009
	TG	HbA1c	0.178*	0.030	0.116	0.18
No diabetes	Chol	HbA1c	−0.070	0.733	0.007	0.975
	LDL-C	HbA1c	−0.096	0.654	−0.091	0.671
	HDL-C	HbA1c	0.393*	0.043	0.511*	0.006
	TG	HbA1c	−0.241	0.217	−0.056	0.776
Prediabetes	Chol	HbA1c	−0.116	0.442	−0.122	0.418
	LDL-C	HbA1c	−0.064	0.678	−0.061	0.693
	HDL-C	HbA1c	−0.143	0.350	−0.084	0.583
	TG	HbA1c	−0.138	0.360	−0.031	0.839
Type 2 diabetes	Chol	HbA1c	−0.220	0.094	−0.209	0.112
	LDL-C	HbA1c	−0.273*	0.038	−0.299*	0.023
	HDL-C	HbA1c	−0.391**	0.002	−0.438**	0.001
	TG	HbA1c	0.148	0.264	0.163	0.217

Table 6. Multinomial logistic regression accuracy scores

Predictors	N/I/O/M/P	W/B
Chol	0.352	0.596
LDL-C	0.342	0.603
HDL-C	0.331	0.577
TG	0.354	0.595
Chol, LDL-C	0.342	0.611
Chol, HDL-C	0.329	0.65
Chol, TG	0.360	0.665
LDL-C, HDL-C	0.376	0.656
LDL-C, TG	0.334	0.659
HDL-C, TG	0.346	0.615
Chol, LDL-C, HDL-C	0.344	0.688
Chol, LDL-C, TG	0.366	0.652
Chol, LDL-C, TG	0.366	0.652
Chol, HDL-C, TG	0.353	0.681
LDL-C, HDL-C, TG	0.360	0.648
Chol, LDL-C, HDL-C, TG	0.368	0.656

5   Discussion

Applying the PBC method (Table 3) shows significant positive correlation with the W class and a significant negative correlation for the B class for Chol, LDL-C and HDL-C. For TG, homoscedasticity was not satisfied for W and B classes. For the specific glucose regulation classes we see significant negative correlations of Chol and LDL-C with P (poor regulation) and with M (medium regulation) for HDL-C. All these results hint at a weak negative correlation between Chol, HDL-C and LDL-C with HbA1c which is the criteria on which these classes are decided.

The univariate logistic regression for binary classification we constructed needed to be highly accurate, sensitive and precise for there to be a strong correlation between the features. As the point-biserial coefficient previously showed, we can see a better classification of the W and B glucose regulation classes with approximately 0.78 and 0.64 PR AUC mean scores for the W and B class respectively which indicates a better correlation, and since there is no intersect of HbA1c values for these classes, the correlation is stronger. This indicates that the lipid profile is more correlated to HbA1c regardless of the diabetes type. The interpretation is based more on the PR curves than on the ROC curves and derives a weak correlation between the glucose regulation classes and the individual features of the lipid profile.

Pearson and Spearman coefficients for linear and non-linear monotonic relationships gave insights into no strong stand-alone correlations between HbA1c levels and the cholesterol features, neither for all patients or divides by the diabetes type. For all patients they showed significant correlations up to absolute 0.3, negative with Chol, LDL-C and HDL-C and positive with TG. For the non-diabetic patients we see a significant positive correlation with HDL-C and the other correlations are not even close to being significant with p-values larger than 0.6. For prediabetic patients there are no significant correlations as the p-values show. For type 2 diabetes there was a significant negative correlation with the LDL-C and with the HDL-C for which the significance is even higher.

The multinomial logistic regression was not expected to give any exceptional results considering the univariate logistic regression did not perform well. This expectation was justified with the accuracy scores which did not surpass 0.7. Again, this confirms the weak correlations.

Regarding the related work shown in Table 7 for type 2 diabetes our results overlapped in the negative correlation with HDL-C, but contrasted with the other papers' findings that the LDL-C was positively correlated whereas we found a negative one. For the TG and Chol they found positive correlations while we found no statistically significant correlations. For prediabetic patients in contrast to Calanna et al. [2] which found negative correlation with HDL-C and a positive correlation with TG, we did not find any significant correlations from our data. The type 1 diabetes related work that is presented in the Table 7 was just an extra on our behalf to see the differences between the diabetes type 1 and 2 correlations, but in our data there are no type 1 diabetic patients.

Table 7. Comparison of correlation results from different sources

Diabetes type	Paper	Chol vs HbA1c	LDL-C vs HbA1c	HDL-C vs HbA1c	TG vs HbA1c
Prediabetes	[2]	None	None	−Negative	+Positive
Type 1	[18]	+Positive	+Positive	None	+Positive
	[16]	+Positive	None	None	+Positive
Type 2	[1]	+Positive	+Positive	None	+Positive
	[3]	+Positive	+Positive	None	+Positive
	[6]	None	None	−Negative	+Positive
	[7]	+Positive	+Positive	None	+Positive
	[8]	+Positive	+Positive	None	+Positive
	[9]	+Positive	+Positive	−Negative	+Positive
	[10]	+Positive	+Positive	−Negative	+Positive
	[11]	+Positive	+Positive	None	+Positive
	[12]	+Positive	+Positive	−Negative	+Positive
	[14]	+Positive	+Positive	−Negative	+Positive
	[15]	+Positive	+Positive	−Negative	+Positive
	[17]	+Positive	+Positive	−Negative	+Positive
	[20]	+Positive	+Positive	None	+Positive
	[23]	+Positive	+Positive	None	+Positive
	[25]	None	+Positive	None	None
No diabetes	This paper	None	None	−Negative	None
Prediabetes		None	None	None	None
Type 2		None	−Negative	−Negative	None
All patients		−Negative	−Negative	−Negative	+Positive

6   Conclusion

The methodologies that were worked through with our data in search for correlations of the lipid profile with HbA1c or the glucose regulation classes gave some significant correlations. Regarding the glucose regulation classes W and B the methods showed statistically significant negative correlations with Chol, HDL-C and LDL-C. When it comes to the correlations of the lipid profile with HbA1c, for all patients there were significant negative correlations with Chol (corr = −0.264, p = 0.002) LDL-C (corr = −0.297, p < 0.001) and HDL-C (corr = −0.28, p = 0.001) and a significant positive correlation with TG (corr = 0.178, p = 0.03). The correlations mentioned are the stronger ones that were found for linear relationships. For non-diabetic patients there was a stronger positive non-linear correlation for HbA1c and HDL-C, (corr = 0.511, p = 0.006) and slightly weaker linear correlation (corr = 0.393, p = 0.043). For prediabetic patients there were no significant correlations. For type 2 diabetes stronger significant negative non-linear correlations were found for HbA1c with LDL-C (corr = −0.299, p = 0.023) and HDL-C (corr = −0.438, p = 0.001). The linear relationships were again, slightly weaker with LDL-C (corr $= -0.273$, p $= 0.038$) and with HDL-C (corr $= -0.391$, p $= 0.002$).

7   Limitations of the Study

The absence of significant correlations could be due to confounding factors. For example, 90% of the patients take statins medications which lowers their lipids and thus influences the resulting correlations.

References

1. Bener, A., Zirie, M., Daghash, M.H., Al-Hamaq, A.O., Daradkeh, G., Rikabi, A.: Lipids, lipoprotein (a) profile and HbA1c among Arabian type 2 diabetic patients (2007)
2. Calanna, S., et al.: Lipid and liver abnormalities in haemoglobin A1c-defined pre-diabetes and type 2 diabetes. Nutr. Metab. Cardiovasc. Dis. **24**(6), 670–676 (2014)
3. Chang, J.B., Chu, N.F., Syu, J.T., Hsieh, A.T., Hung, Y.R.: Advanced glycation end products (AGEs) in relation to atherosclerotic lipid profiles in middle-aged and elderly diabetic patients. Lipids Health Dis. **10**(1), 228 (2011). https://doi.org/10.1186/1476-511X-10-228
4. Chen, P., Popovich, P.: Quantitative Applications in the Social Sciences: Correlation, vol. 10, p. 9781412983808. SAGE Publications Ltd., Thousand Oaks (2002)
5. Davis, J., Goadrich, M.: The relationship between Precision-Recall and ROC curves. In: Proceedings of the 23rd International Conference on Machine Learning, pp. 233–240 (2006)
6. Drexel, H., et al.: Is atherosclerosis in diabetes and impaired fasting glucose driven by elevated LDL cholesterol or by decreased HDL cholesterol? Diab. care **28**(1), 101–107 (2005)
7. Eftekharian, M.M., Karimi, J., Safe, M., Sadeghian, A., Borzooei, S., Siahpoushi, E.: Investigation of the correlation between some immune system and biochemical indicators in patients with type 2 diabetes. Hum. Antibodies **24**(1–2), 25–31 (2016)
8. Ghari Arab, A., Zahedi, M., Kazemi Nejad, V., Sanagoo, A., Azimi, M.: Correlation between hemoglobin A1c and serum lipid profile in type 2 diabetic patients referred to the diabetes clinic in Gorgan, Iran. J. Clin. Basic Res. **2**(1), 26–31 (2018)
9. Grant, T., et al.: Community-based screening for cardiovascular disease and diabetes using HbA1c. Am. J. Prev. Med. **26**(4), 271–275 (2004)
10. Hu, Y., Liu, W., Huang, R., Zhang, X.: Postchallenge plasma glucose excursions, carotid intima-media thickness, and risk factors for atherosclerosis in Chinese population with type 2 diabetes. Atherosclerosis **210**(1), 302–306 (2010)
11. Hussain, A., Ali, I., Ijaz, M., Rahim, A.: Correlation between hemoglobin A1c and serum lipid profile in Afghani patients with type 2 diabetes: hemoglobin A1c prognosticates dyslipidemia. Ther. Adv. Endocrinol. Metab. **8**(4), 51–57 (2017)
12. Hwang, Y.C., Ahn, H.Y., Park, S.W., Park, C.Y.: Apolipoprotein B and non-HDL cholesterol are more powerful predictors for incident type 2 diabetes than fasting glucose or glycated hemoglobin in subjects with normal glucose tolerance: a 3.3-year retrospective longitudinal study. Acta Diabetol. **51**(6), 941–946 (2014)

13. Innovation Dooel: Glyco project - measure ECG and glucose levels with a small, non-invasive, wearable monitor (2019). http://glyco.innovation.com.mk/. Project partially funded by Fund of Innovations and Technical Development, North Macedonia

14. Khan, H., Sobki, S., Khan, S.: Association between glycaemic control and serum lipids profile in type 2 diabetic patients: HbA 1c predicts dyslipidaemia. Clin. Exp. Med. **7**(1), 24–29 (2007). https://doi.org/10.1007/s10238-007-0121-3

15. Khan, H.A.: Clinical significance of HbA 1c as a marker of circulating lipids in male and female type 2 diabetic patients. Acta Diabetol. **44**(4), 193–200 (2007). https://doi.org/10.1007/s00592-007-0003-x

16. Kim, S.H., et al.: Serum lipid profiles and glycemic control in adolescents and young adults with type 1 diabetes mellitus. Ann. Pediatr. Endocrinol. Metab. **19**(4), 191 (2014)

17. Lopes-Virella, M.F., Wohltmann, H.J., Mayfield, R.K., Loadholt, C., Colwell, J.A.: Effect of metabolic control on lipid, lipoprotein, and apolipoprotein levels in 55 insulin-dependent diabetic patients: a longitudinal study. Diabetes **32**(1), 20–25 (1983)

18. Prado, M.M., Carrizo, T., Abregú, A.V., Meroño, T.: Non-HDL-cholesterol and C-reactive protein in children and adolescents with type 1 diabetes. J. Pediatr. Endocrinol. Metab. **30**(3), 285–288 (2017)

19. Raghavan, V., Bollmann, P., Jung, G.S.: A critical investigation of recall and precision as measures of retrieval system performance. ACM Trans. Inf. Syst. (TOIS) **7**(3), 205–229 (1989)

20. Reddy, S., Meera, S., William, E., Kumar, J.: Correlation between glycemic control and lipid profile in type 2 diabetic patients: HbA1c as an indirect indicator of dyslipidemia. Age **53**, 10–50 (2014)

21. Saito, T., Rehmsmeier, M.: The precision-recall plot is more informative than the ROC plot when evaluating binary classifiers on imbalanced datasets. PloSone **10**(3), e0118432 (2015)

22. Sheskin, D.J.: Handbook of Parametric and Nonparametric Statistical Procedures. CRC Press, Boca Raton (2003)

23. VinodMahato, R., et al.: Association between glycaemic control and serum lipid profile in type 2 diabetic patients: glycated haemoglobin as a dual biomarker (2011)

24. Vishinov, I., Gusev, M., Poposka, L., Vavlukis, M.: Correlating the cholesterol levels to glucose for men and women. Submitted to the CiiT 2020, Ss. Cyril and Methodius University, Faculty of Computer Science and Engineering (2020)

25. Yan, Z., Liu, Y., Huang, H.: Association of glycosylated hemoglobin level with lipid ratio and individual lipids in type 2 diabetic patients. Asian Pac. J. Trop. Med. **5**(6), 469–471 (2012)

Author Index

Alcaraz, Salvador 111
Anchev, Nenad 1

Bakeva, Verica 162
Beredimas, Nikolaos 28
Bikov, Dushan 138

Chorbev, Ivan 98
Chouvarda, Ioanna 28
Corbev, Ivan 28

Despotovski, Aleksandar 56
Despotovski, Filip 56
Dobreva, Jovana 87

Filiposka, Sonja 111, 174

Gavrilov, Goce 202
Gievska, Sonja 42
Gilly, Katja 111
Gjorgiev, Laze 42
Guliashki, Vassil 71
Gusev, Marjan 217

Jakimovski, Boro 1, 28
Jofche, Nasi 87
Jolevski, Ilija 28
Jovanovik, Milos 87

Kalajdjieski, Jovan 15
Kilintzis, Vassilis 28
Kitanovski, Dimitar 98
Kjorveziroski, Vojdan 174
Kon-Popovska, Margita 189
Korunoski, Mladen 15
Kulakov, Andrea 56

Lameski, Jane 56
Lameski, Petre 56

Madevska Bogdanova, Ana 64
Maglaveras, Nicos 28
Mankolli, Emiliano 71
Markoski, Marko 64
Mechkaroska, Daniela 162
Mishev, Anastas 174
Mourouzis, Alexandros 28

Nikolovski, Vlatko 98

Pashinska, Maria 138
Poposka, Lidija 217
Popovska-Mitrovikj, Aleksandra 162

Raça, Vigan 189
Roig, Pedro Juan 111

Simov, Orce 202
Snezana, Savoska 28
Spasov, Dejan 127, 153
Spiridonov, Vlado 1
Stoimenov, Leonid 189
Stojkoska, Biljana Risteska 15

Trajanov, Dimitar 87, 98
Trajkovik, Vladimir 28, 202
Trivodaliev, Kire 15

Vavlukis, Marija 217
Velinov, Goran 1, 189
Veljković, Nataša 189
Vishinov, Ilija 217

Zdravevski, Eftim 56

Printed in the United States
By Bookmasters